MICRO and NANO TECHNIQUES for the HANDLING of BIOLOGICAL SAMPLES

Edited by
Jaime Castillo-León
Winnie Edith Svendsen
Maria Dimaki

CRC Press
Taylor & Francis Group
Boca Raton London New York

CRC Press is an imprint of the
Taylor & Francis Group, an **informa** business

CRC Press
Taylor & Francis Group
6000 Broken Sound Parkway NW, Suite 300
Boca Raton, FL 33487-2742

First issued in paperback 2018

© 2012 by Taylor & Francis Group, LLC
CRC Press is an imprint of Taylor & Francis Group, an Informa business

No claim to original U.S. Government works

ISBN-13: 978-1-4398-2743-7 (hbk)
ISBN-13: 978-1-138-38199-5 (pbk)

Visit the Taylor & Francis Web site at
http://www.taylorandfrancis.com

and the CRC Press Web site at
http://www.crcpress.com

To our families

Contents

Preface

The Nano-Bio Integrated Systems group (NaBIS) at DTU Nanotech was created in 2006 by Associate Professor Winnie E. Svendsen with a single activity as the aim, namely, the fabrication of a chromosome total analysis system. Starting with two members, the group quickly evolved into one of the largest groups at the institute, currently consisting of 18 people and participating in more than six different national and international research programs.

This book is the direct result of a report written for the EU-funded project BeNATURAL by Assistant Professor Jaime Castillo-León. The EU commissioner responsible for the project encouraged the writing of a review article based on that report. Jaime enlisted the help of group leader Associate Professor Winnie E. Svendsen and Associate Professor Maria Dimaki, and in 2009 a review article appeared in *Integrative Biology* (Castillo et al., 2009, pp. 30–42) that dealt with the manipulation of biological samples using micro- and nanotechniques.

This article received some attention and resulted in being asked if we would be interested in expanding it into a book. Well aware of the effort required, we agreed and enlisted the help of a group of experts at the institute and abroad, each a pioneer in his fields. The book proposal received some very positive reviews and suddenly it was a reality.

The road to this now-finished work was not easy and involved a lot of frustrations and hard effort. However, we are quite pleased with the result, which we feel will appeal to a broad audience, as we strived to achieve a balance between simplicity for the beginner and hardcore theory for the more advanced readers.

MATLAB® is a registered trademark of The MathWorks, Inc. For product information, please contact:

The MathWorks, Inc.
3 Apple Hill Drive
Natick, MA 01760-2098 USA
Tel: 508 647 7000
Fax: 508-647-7001
E-mail: info@mathworks.com
Web: www.mathworks.com

Acknowledgments

The editors would like to thank all the authors who agreed to contribute to this book. We thank Associate Professor Zachary James Davis from Denmark Technical University (DTU) Nanotech, Associate Professor Kirstine Berg-Sørensen from DTU Physics, Dr. Stefano Oberti and Professor Jürg Dual from the Institute of Mechanical Systems (ETHZ, Switzerland), Associate Professor Peter Bøggild from DTU Nanotech, and Associate Professor Fridolin Okkels from DTU Nanotech who have undertaken the enormous task of summarizing their field of expertise in a limited amount of pages. It wasn't an easy task, but it is one they have all done very well.

We would also like to thank Associate Professor Mikkel Fougt Hansen from DTU Nanotech for many useful discussions and proofreading of the theoretical treatment of magnetic forces on particles, a field of research for which remarkably varying equations can be found in the literature.

We would further like to thank all the members of the Nano-Bio Integrated Systems (NaBIS) group at DTU Nanotech not only for their scientific contribution to this book but also for their understanding and patience when the writing process took us away from the everyday activities of the group.

Finally, we would like to thank our families for their continuing support during the sometimes stressful periods that arose in the last year that the book was in preparation.

The Authors

Jaime Castillo–León graduated from the Industrial University of Santander (Bucaramanga, Colombia) with a BSc in chemistry. He received his PhD degree in 2005 at the Department of Biotechnology, Lund University, Sweden. His PhD work involved the fabrication of electrochemical biosensors for detection of compounds of biomedical importance using cellular models. He currently holds a position as assistant professor at the Department of Micro- and Nanotechnology, DTU Nanotech, Technical University of Denmark. His research focuses on micro- and nanotechnologies for the development of biosensing devices for biomedical applications. A strong focus is currently set on manipulation, characterization, and integration of biological nanotubes and nanofibers with micro- and nanostructures for the development of bioelectronic sensing devices and drug delivery systems.

Maria Dimaki graduated among the top 2% of her year with an MEng from the Department of Electrical and Computer Engineering, National Technical University of Athens in 2000 and first in her year with distinction with an MSc degree in engineering and physical science in medicine in 2001 from the Department of Bioengineering, Imperial College, London. Dr. Dimaki then joined the Department of Micro- and Nanotechnology at the Technical University of Denmark, where she received her PhD degree in 2005 after having worked with the assembly of carbon nanotube devices using dielectrophoresis. Dr. Dimaki has remained with the department since then, and is currently an associate professor within the Nano-Bio Integrated Systems group (NaBIS). Her research interests mainly involve the development of micro- and nanoelectrodes in microfluidic systems for electrophysiological and electrochemical measurements on neuron cultures by use of standard microfabrication processes. However, due to her expertise and interest in the matter, Dr. Dimaki is also involved in theoretical simulations of dielectrophoresis on biological particles using COMSOL Multiphysics and MATLAB®.

Winnie Svendsen received her BSc degree in 1992 and MSc degree in physics in 1993 from the University College Dublin, Ireland; here, she received the EOLAS applied research award for excellent research. Her PhD was from Copenhagen University and the National Laboratory for Sustainable Energy (RISØ), and was finalized in 1996. In 1996 she accepted a postdoctoral position at the Max Planck Institute for plasma physics. In 1998 Dr. Svendsen received talent stipend from SNF (now FNU) and the prestigious Curie stipend from Copenhagen University to establish a research group to design a hyperpolarized gas set-up for use in medical lung diagnostic. In connection with this project she received funding from the European Fifth Framework to organize a workshop on hyperpolarized gasses. In 1999 she was appointed associate professor at Copenhagen University. She was the cofounder of the company XeHe Hypol (APS). Since 2000 Dr. Svendsen has been employed as associated professor at DTU. In 2006 she established her own research group, Nano-Bio Integrated Systems (NaBIS).

Zachary J. Davis got his MSc in applied physics in 1999 and his PhD in 2003 in electrical engineering from the Technical University of Denmark. He became an assistant professor at DTU Nanotech in 2004. In 2006 he was appointed as associate professor at DTU Nanotech. His research area focuses on dynamic nano mechanical systems toward biochemical detection. His current work focuses on the development of bulk mode resonators based sensors for gas and liquid based sensing. Dr. Davis is the head of the Dynamics NEMS group at DTU Nanotech.

Kirstine Berg-Sørensen is associate professor in biological physics at the Department of Physics, Technical University of Denmark. She was educated at the University of Aarhus, with a 1-year stay at the École Normale Supérieure in Paris, France during her PhD studies. Her PhD work and first postdoctoral experiences were concentrated on quantum mechanical and semiclassical models for laser cooling and the physics of cold atoms, as a postdoctoral fellow with Lene Hau, Rowland Institute for Science, and Chris Pethick, Nordic Institute for Theoretical Physics (NORDITA). Later, her research activities became focused on the physics of biological systems, as her interest in this topic was triggered by some of the first reports on the use of optical tweezers in single molecule biophysics. Dr. Berg-Sørensen own research in the field of biological physics has been concentrated on simple models and precise data analysis for single molecule experiments, under both *in vitro* and *in vivo* conditions, with the main emphasis on optical trapping by optical tweezers and optical stretchers.

Jürg Dual has been professor of mechanics and experimental dynamics in the Center of Mechanics of the Institute of Mechanical Systems at the Swiss Federal Institute of Technology (ETH) in Zurich since October 1, 1998. He is president of the University Assembly of ETH Zurich and continues to be a member. Dr. Dual studied mechanical engineering at the ETH Zurich. He then spent 2 years on a Fulbright grant at the University of California–Berkeley, where he graduated with a MS and a MEng degree in mechanical engineering. He then received his Dr. Sc. Techn. degree at the ETH Zurich under the guidance of Prof. Dr. M. Sayir at the Institute of Mechanics. For his dissertation he was awarded the Latsis Prize of the ETH Zurich in 1989. After one year as visiting assistant professor at Cornell University, Ithaca, New York, he returned to the ETH Zurich as assistant professor. He is a fellow of the American Society of Mechanical Engineers (ASME) and honorary member of the German Association for Materials Research and Testing. His research focuses on wave propagation and vibrations in solids as well as micro- and nanosystem technology. In particular he is interested in both basic research and applications in the area of sensors (viscometry), nondestructive testing, mechanical characterization of microstructures and gravitational interaction of vibrating systems. In his research, experimentation is central, but must always be embedded in corresponding analytical and numerical modeling. As mechanics is a very basic science, it is particularly attractive for him to interact with neighboring disciplines such as bioengineering, materials science, or micro- and nanosystem technology.

Stefano Oberti earned his degree in mechanical engineering from the Swiss Federal Institute of Technology in 2004. During his studies he specialized in micro- and nanotechnology, as well as control systems. After completion of his diploma thesis at the IBM Zurich Research Laboratory on distance measurements in liquid environment based on the force generated by the fluid squeezed between two surfaces, he entered the field of acoustic particle manipulation. Until 2009, when he received his PhD from the Swiss Federal Institute of Technology, he investigated manipulation of suspended micron-sized particles by means of acoustic fields within micromachined fluidic systems.

Peter Bøggild is an associate professor heading the nanointegration research group in the Department for Micro- and Nanotechnology (DTU Nanotech) at the Technical University of Denmark. He received his PhD degree at Copenhagen University in 1998 in the field of experimental low temperature physics, with the title of the thesis "Electron Transport in Open Quantum Dots." Dr. Bøggild was employed as assistant research professor at MIC from 1998 to 2001, investigating applications for micro-scale four-point probes for measurements in ultra-high vacuum on self-assembled polymeric monolayers and on metallic thin films. In 2001 he was appointed associate professor, and since then built a research group concerned with the development of new nanoscale tools for manipulation and characterization of nanostructures, including micro- and nanoscale four-point probes, microtweezers for 3D nanorobotics, and fast prototyping technology, as well as integration of carbon nanotubes and silicon nanowires with standard silicon microfabrication techniques. Recently, the research is focusing on integration of graphene with silicon devices.

Fridolin Okkels graduated from Copenhagen University with an MSc in physics. He received his PhD in 2001 from the same university. His PhD work involved the creation and interaction of self-organized structures in turbulence. From 2001 to 2003 he did postdoctoral work at the École Normale Supérieure, ENS, and the École Supérieure de Physique et de Chimie Industrielles, ESPCI, in Paris. Dr. Okkels has been working at DTU Nanotech since 2003, initially as associate professor. In 2007 he was appointed as associate professor. His research area focuses on structural optimization, microfluidics, lab-on-a-chip systems, hydrodynamics, thermodynamics, energy systems, and chemical and biological microreactors. He is the leader of the Theoretical Microsystems Optimization group.

Contributors

Kirstine Berg-Sørensen
Department of Physics
Technical University of Denmark
Lyngby, Denmark

Peter Bøggild
Department of Micro- and
    Nanotechnology
Technical University of Denmark
Lyngby, Denmark

Jaime Castillo-León
Department of Micro- and
    Nanotechnology
Technical University of Denmark
Lyngby, Denmark

Zachary James Davis
Department of Micro- and
    Nanotechnology
Technical University of Denmark
Lyngby, Denmark

Maria Dimaki
Department of Micro- and
    Nanotechnology
Technical University of Denmark
Lyngby, Denmark

Jürg Dual
Institute of Mechanical Systems
ETH Zentrum
Zurich, Switzerland

Stefano Oberti
Institute of Mechanical Systems
ETH Zentrum
Zurich, Switzerland

Fridolin Okkels
Department of Micro- and
    Nanotechnology
Technical University of Denmark
Lyngby, Denmark

Winnie E. Svendsen
Department of Micro- and
    Nanotechnology
Technical University of Denmark
Lyngby, Denmark

1 Interfacing Biological Material with Micro- and Nanodevices

Jaime Castillo-León and Winnie E. Svendsen
Department of Micro- and Nanotechnology,
Technical University of Denmark, Lyngby

CONTENTS

1.1 INTRODUCTION

Life is possible thanks to the perfect harmony of thousands of different molecules, proteins, cells, tissues, etc., working in a precise and coordinated way and allowing us to perform all our everyday activities. In order for this huge ensemble of biological structures to continue working without problems, each biological component requires specific conditions to ensure for itself enough energy and to protect it from denaturalizing elements that could create problems, causing the machinery to fail. Advances in fields such as microscopy, biology, nanotechnology, and medicine offer us the opportunity to find and evaluate these failures by collecting, isolating, and studying some of the biological structures responsible for vital functions. The idea is to extract as much information as possible from them. In this way, we can gain a better understanding of these biological structures: their composition, their interaction with other similar structures or with different materials, and their reaction to external stimulus and to varying environmental conditions. It would, furthermore, allow us to predict, for instance, how cells will respond to the presence of a specific drug

1

compound. For this, biological samples often need to be manipulated in such a way that their natural environment (temperature, ionic strength, shape, pressure, etc.) is preserved to as large a degree as possible in order to avoid changes in the sample that will be reflected in inaccurate results. This chapter presents several aspects regarding the challenges involved in the manipulation of biological samples using different techniques in 2D and 3D. The consequences of going down in size for the manipulation techniques are also discussed.

1.2 LINK BETWEEN THE MACRO, MICRO, AND NANOWORLD

One of the biggest and more critical challenges for researchers in biomedical fields during recent years has been how to manipulate and integrate biomaterial into various devices, such as microelectrodes, sensor platforms, or microfluidic chips. The size, structural composition, and the manner in which these biomaterials react under external forces (mechanical, electrical, magnetic, etc.) determine the success of an experiment. Fortunately, several techniques have, during the last decades, appeared to help researchers treat these samples in a correct way, rendering it possible to maintain them in an environment that preserves their properties.

Some of the methods or techniques involve a direct contact between the manipulator (microgrippers, atomic force microscopy) and the biological object, whereas others involve the application of external forces (dielectrophoresis [DEP], optical tweezers) that do not come in direct contact with the sample. Depending on the choice of the method, the challenges involved in the manipulation of the biomaterial change. However, not only does the technique used to manipulate the biomaterial determine the challenges to overcome but the medium in which the sample is present will also influence the degree of difficulty when controlling the samples. Different strategies need to be employed depending on whether the biological sample is in air, liquid, or a vacuum environment.

The manipulation of micrometer- and nanometer-scale biomaterial and their integration in different devices has become a critical issue in the bio-nanotechnology field during the last decade (Amato, 2005; Desai et al., 2007; Laval et al., 2000). The linking of our macroscopic world to the nanoscopic one of single molecules, nanoparticles, and functional nanostructures is a technological challenge (Bolshakova et al., 2010; Cohen, 2007; Fukuda et al., 2003; O'Malley, 2007; Rubio-Sierra et al., 2005; Sahu et al., 2010). Various techniques have emerged to help scientists overcome the obstacles of size and precision when interacting with tiny objects such as cells, proteins, viruses, and self-assembled nanotubes, among others.

The selection of the right technique to manipulate the biological component will depend on several aspects such as its size and shape, or the medium, either dry or wet, in which the component exists (Scheme 1.1). If some of the biological samples are in the nanometer scale (e.g., hepatitis virus, antibodies, ribosomes), the manipulation of these entities includes, first, finding these objects, tracking them, and finally moving them. Two-dimensional nanomanipulation using scanning probe microscopy renders it possible to move nanocomponents along a surface with nanometer-level precision. However, the transfer of nanocomponents from a "source substrate" to a "target substrate," such as picking a nanocomponent from a solution and placing it between two

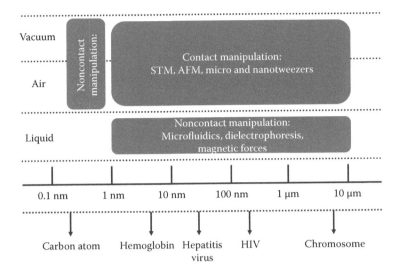

SCHEME 1.1 The different strategies for manipulation are determined by the environment (air, liquid, or vacuum), which is decided by the properties and size of the objects, and observation methods. (Modified from Fukuda, T., Arai, F., and Dong, L., Assembly of nanodevices with carbon nanotubes through nanorobotic manipulations, *Proceedings of the IEEE*, 91, 1803–1818 © [2003])

electrodes or mounting it as the extension of a scanning probe tip, is difficult without a full 3D monitoring and manipulation capability (Castillo et al., 2009).

While 3D manipulation is essential to transferring biological samples to a specific location, it is instructive to review the established methods of visualization and nanomanipulation. Typically, the type of microscope, sample size, and environment in which the sample exists determine the possibilities and limitations in the case of 2D and 3D manipulation, as presented in Scheme 1.2. For instance, scanning probe microscopes, scanning tunneling microscopes, and atomic force microscopes (AFMs) offer the best resolution and precision, but in some cases only work in 2D, while electron microscopes provide more possibilities to manipulate biological objects in 3D. Atomic force microscopy provides a powerful tool for manipulation of biological samples in a liquid environment, while in most cases the use of scanning electron microscopes with biological samples requires ultrahigh vacuum.

In the case of optical microscopes (OMs), they can be combined with manipulation techniques such as microfluidics for the visualization and trapping of single cells (Sott et al., 2008; Spegel et al., 2008), dielectrophoresis for the study of parasites (Dalton et al., 2006), or acoustic waves for the manipulation of neuronal cells (Muratore et al., 2009). Unfortunately, their use is limited by lens aberrations, diffraction effects, and samples with sizes smaller than the wavelength of visible light. However, advances in the development of scanning near-field optical microscopes (SNOMs) allow us to overcome this limitation (De Serio et al., 2003). The SNOM was combined with manipulation techniques such as AFM for the visualization and manipulation of individual proteins (Muller et al., 2002) and chromosomes

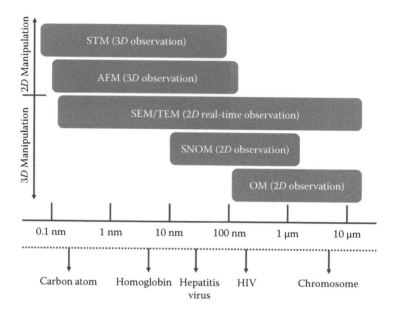

SCHEME 1.2   Microscopes, environments, and strategies for manipulation of biological samples. Scanning tunneling microscope (STM), atomic force microscopy (AFM), scanning electron microscopy (SEM), transmission electron microscopy (TEM), optical microscope (OM), and scanning near-field optical microscope (SNOM). (Modified from Fukuda, T., Arai, F., and Dong, L., Assembly of nanodevices with carbon nanotubes through nanorobotic manipulations, *Proceedings of the IEEE*, 91, 1803–1818 © [2003])

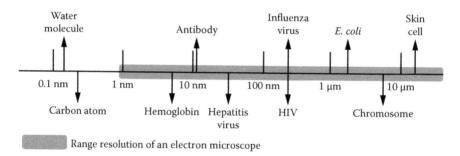

FIGURE 1.1   Relative sizes of various biological objects and the range of a scanning force microscope (SEM).

(Oberringer et al., 2003), and with optical tweezers and microfluidics for single-cell studies (Ramser and Hanstorp, 2010).

1.3   EFFECTS OF SCALING DOWN

One of the big challenges for researchers aiming to manipulate biological micro- or nanostructures is to understand and exploit the changes in behavior that occur when going down in size. Figure 1.1 indicates the relative scale of some of the biological samples that will be discussed in the following chapters. Effects that are negligible at the macroscopic scale become dominant at the micro- and nanometer scales. This situation forces us to modify our way of thinking by taking into account the new effects that dominate at these scales. An example of this scaling effect can be found in microfluidics. When decreasing the size, gravity no longer plays a major role in a microchannel. On the other hand, forces that are insignificant at the macroscale are now dominating, such as surface tension (Castillo et al., 2009).

It is thus necessary to be aware of and to understand the scaling laws and their effects in different fields such as mechanics, electromagnetism, fluids, optics, thermodynamics, and quantum effects. This will facilitate the understanding of the transition from the macroscopic to the micro- and nanoscopic dimensions. A more detail description of these scaling laws is beyond the scope of this chapter; there are several reviews that deal with the topic of scaling (Abbot et al., 2007; Cugat et al., 2003; Dong and Nelson, 2007; Duvivier et al., 2008; Purcell, 1977; Trimmer, 1989; Wautelet, 2001).

1.4 CHALLENGES IN THE MANIPULATION OF BIOLOGICAL SAMPLES

"Biological samples" is a broad term covering anything from DNA strings with a diameter of about 2 nm; bacteria and viruses in the submicron range; and nerve cells, self-assembled peptide tubes, or particles that can be hundreds of micrometer long. Here the focus is on the nano- and microscale, roughly from tens of nanometers to tens of micrometers. The biomanipulation of these samples is complicated by several issues.

1.4.1 FRAGILITY OF BIOMATERIALS

Biological materials are soft, fragile, easily contaminated, heat sensitive, and sometimes even alive, which puts severe limitations on the operation and design of the tools not encountered when manipulating tough, rigid, inorganic material such as carbon nanotubes (Andersen et al., 2009; Park et al., 2010). This calls for nontoxic materials and gripper layouts that do not heat up or otherwise interfere with the objects to be manipulated, except for the application of as little force as possible. Integrated force feedback to adjust the applied force accurately is another desirable feature to avoid mechanical forces damaging the samples (Johnstone and Parmaswaran, 2004).

1.4.2 WET ENVIRONMENT

While there are many reports of successful contact manipulation of objects in air as well as in vacuum, there are considerably fewer in the liquid environment. Electrothermal (Liu et al., 2008; Sahu et al., 2010) and electrostatic (Lu and Kim, 2006; Luo et al., 2005) actuators that work excellently in scanning electron microscopes experience problems in a saline solution, which is both electrically and thermally conductive. The temperature generated by metallic or silicon electrothermal actuators can be substantial—up to hundreds of degrees (Rubio et al., 2009)—and this can destroy the samples to be manipulated. In aqueous solutions, the gripper arms will be cooled as the water will transport the heat away from the gripper, heating up the solution instead. Electrostatic actuators work by applying a voltage ranging from a few to hundreds of volts across small gaps, thus creating large electrical fields. Ionic current flow can make electrostatic actuation impractical, and hydrolysis occurs already at 1–2 V in ionic solution (Rubio et al., 2009). One strategy is to operate the actuator *outside* the liquid and transfer the mechanical force through the liquid boundary and into the water so that only a smaller part of the gripper is actually immersed (Mølhave et al., 2006); another is to avoid voltages larger than a few volts or use alternating current (AC) voltage at high frequencies, which inhibits electrolysis (Marie et al., 2005).

Noncontact manipulation is always done inside a liquid, and this also has its fair share of problems. These will be presented in detail in Chapters 3, 4, 5, 7, and 8.

1.4.3 Limited Visibility and Accessibility

Aqueous solutions and biomaterials generally preclude the use of electron microscopy because of vacuum requirements, unless sealed *in situ* liquid cells with windows for the transmission of electrons are used; in this case, biological cells can be monitored with nanoresolution in a transmission electron microscope (TEM) (Landau and Lifshitz, 1960). There is presently no known way to perform active manipulation in a closed liquid cell with the extremely narrow space requirements of TEM. This leaves optical microscopes as the only viable option for manipulation of biological specimens. Microfabricated tools are typically much more bulky than glass capillaries and require the longer working distances of the optical microscopes to fit in, which leads to challenges in achieving sufficiently high microscopic resolution. As a result, it can be difficult to arrange the gripper, the object to be manipulated, and the microscope so that the resolution, the space within to move the manipulation tool, and maintaining a free line of sight allow meaningful experiments to be carried out. Similar problems, though less serious, are encountered when electrical fields are used for the manipulation; here the space requirements arise from electrical and fluidic connectors. This is one of the reasons that many experimental demonstrations never reach beyond "proof-of-concept" status.

1.4.4 Interaction Forces between the Biological Material and the Manipulation Tool

As mentioned earlier, biological materials are in most cases soft and fragile. They can easily be irreversibly deformed by the various tools used to manipulate or image them. Dynamic AFM is a technique widely used for the imaging and manipulation of biological materials. For this, a sharp tip interacts with the sample through average and peak forces during the scanning process. Peak forces are those on the order of nanonewtons, and they play an important role in scanning biological samples. Figure 1.2 gives an example of the use of the AFM for the manipulation of biological samples. This system is called the *nanomanipulator* and has been employed for the manipulation and investigation of biological

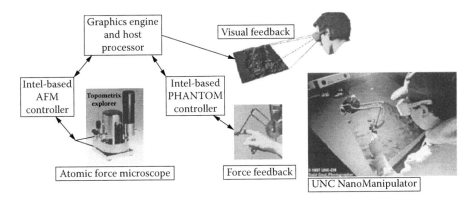

FIGURE 1.2 Schematics of the nanomanipulator system for the modification of molecular structures using atomic force microscopy (AFM). (Reprinted from Guthold, M., Falvo, M., Matthews, W. G., Paulson, S., Mullin, J., Lord, S., Erie, D., Washburn, S., Superfine, R., Brooks, F. P., and Taylor, R. M. (1999a), *Journal of Molecular Graphics & Modelling*, 17, 187–197. With permission from Elsevier.)

samples such as fibrin, DNA, adenovirus, and tobacco mosaic virus (Guthold et al., 1999a; Guthold et al., 1999b).

Another method used for the manipulation of biological samples involving a direct interaction is "pick and place." A detailed explanation of this technique will be given in Chapter 6. Pick and place was used during the manipulation of self-assembled peptide nanotubes in order to study their structural and electrical properties. A metal wire is utilized to lift and move an individual peptide nanotube lying on a polymer nanoforest, as shown in Figure 1.3. The strong adhesive forces between the peptide nanotube and the polymer nanoforest did not allow the whole peptide nanotube to be lifted. As displayed in Figure 1.3, the peptide tube broke when the titanium wire lifted it, and only a small part of the tube was moved. A modification of the properties of the polymer nanoforest is required in order to decrease the interaction forces between the tube and the nanoforest to facilitate the displacement of the peptide tube.

These two examples highlight the importance of the interaction forces between the manipulation tool and the biological sample. If these forces are too strong, the handling of the sample could be difficult, resulting in its destruction.

1.5 SELECTION OF THE MANIPULATION TECHNIQUE

It is obvious that the choice of the best manipulation method requires the consideration of multiple factors as was discussed earlier. Unfortunately in almost all the cases, due to the technical requirements, the sample cannot be kept in its natural environment, or the manipulator would suffer some damage.

The challenge is then to develop new techniques, rendering it possible to manipulate biological entities without changing their structure. In this way, the information obtained from them will be more trustworthy and precise.

As presented in the following chapters, several steps have already been taken in this direction, such as in the case of environmental scanning electron

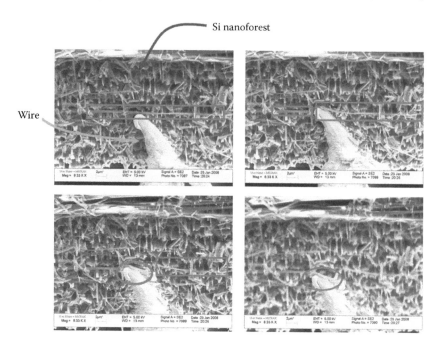

FIGURE 1.3   Pick-and-place of a self-assembled peptide nanotube. A metal wired is used to lift and move and individual self-assembled peptide nanotube lying on a polymer nanoforest (top left). The strong adhesive forces did not allow the whole nanotube to be lifted. The peptide tube broke when the metal wire lifted it (top right and bottem left). Only a small part of the nanotube was moved (bottom right).

microscopy (ESEM), the integration of microfluidics with DEP, or magnetic fields. Nevertheless, several issues still need to be resolved. With the amount of research carried out in this field, it is certain that more methods will be developed in the future so that we can eventually talk about manipulation techniques for biological samples that will interact with them without damaging or changing their natural structures.

With the new and rapid development of microfluidics, manipulating biological material has reached new horizons. By using microfluidics to maintain natural conditions for living organisms, it has been possible to manipulate biomaterials under their natural environments. Combining microfluidics with other manipulation techniques has given researchers a very strong tool for real biomanipulation. The topic of microfluidics is discussed in detail in Chapter 8, and the combination of microfluidics and other manipulation techniques is discussed in the relevant chapters.

1.6   ABOUT THIS BOOK

Chapters 3, 4, 5, and 7 seek to present and discuss the more important techniques used for the handling of biological samples in order to characterize, move, and extract information from these material without damaging or altering their natural state. In this way, the study of biological samples will be as close to the real *in vivo* situation as we can obtain in a research lab.

This chapter begins by pointing out the challenges when working with samples at the micro- and nanoscale, the link between the manipulation tool and the biological sample, as well as the considerations when working with biological material regarding its shape, composition, and the environment in which the sample is present. Chapter 1 also invites the reader to consider the effects of scaling down in size and to be aware of how some properties that play an important role at macroscale became negligible at the micro- and nanoscale.

Subsequent chapters discuss manipulation techniques involving direct contact with the sample such as AFM or micro- and nanotweezers for 2D and 3D manipulation. Noncontact techniques involving the use of external forces are presented in chapters describing methods such as DEP, optical tweezers, magnetic forces, and microfluidics.

These chapters start with a basic theoretical description of each technique, including equations and principles, in order to provide the reader with a basic knowledge that facilitates the understanding of how the different techniques could be used together with biological material. Next, relevant examples are described, discussing the advantages and challenges that arise when manipulating biological material.

This book is aimed at academicians (biomedical engineering, chemical engineering, chemistry, and biochemistry), physicians, and scientists working in the areas of biomaterials, cell handling, bionanotechnology, drug delivery, tissue engineering, and regenerative medicine.

REFERENCES

Abbot, J. J., Nagy, Z., Beyeler, F., and Nelson, B. J. (2007). Robotics in the small. Part I: Microrobotics. *IEEE Robotics and Automation Magazine,* 14, 92–103.

Amato, I. (2005). Nanotechnologists seek biological niches. *Cell* 123, 967–970.

Andersen, K. N., Petersen, D. H., Carlson, K., Molhave, K., Sardan, O., Horsewell, A., Eichhorn, V., Fatikow, S., and Boggild, P. (2009). Multimodal electrothermal silicon microgrippers for nanotube manipulation. *IEEE Transactions on Nanotechnology,* 8, 76–85.

Bolshakova, A. V., Vorobyova, E. A., and Yaminsky, I. V. (2010). Atomic force microscopy studies of living bacterial cells in native soil and permafrost. *Materials Science and Engineering B—Advanced Functional Solid-State Materials,* 169, 33–35.

Castillo, J., Dimaki, M., and Svendsen, W. E. (2009). Manipulation of biological samples using micro- and nano techniques. *Integrative Biology,* 1, 30–42.

Cohen, W. (2007). Manipulation of the very small. *A Computer Scientist's Guide to Cell Biology.* New York, Springer US.

Cugat, O., Delamare, J., and Reyne, G. (2003). Magnetic micro-actuators and systems (MAGMAS). *IEEE Transactions on Magnetics,* 39, 3607–3612.

Dalton, C., Goater, A. D., and Smith, H. V. (2006). Fertilization state of *Ascaris suum* determined by electrorotation. *Journal of Helminthology,* 80, 25–31.

De Serio, M., Zenobi, R., and Deckert, V. (2003). Looking at the nanoscale: Scanning near-field optical microscopy. *TrAC Trends in Analytical Chemistry,* 22, 70–77.

Desai, J. P., Pillarisetti, A., and Brooks, A. D. (2007). Engineering approaches to biomanipulation. *Annual Review of Biomedical Engineering,* 9, 35–53.

Dong, L. and Nelson, B. J. (2007). Robotics in the small. Part II. Nanorobotics. *IEEE Robotics and Automation Magazine,* 14, 111–121.

Duvivier, D., Van Overschelde, O., and Wautelet, M. (2008). Nanogeometry. *European Journal of Physics,* 29, 467–474.

Fukuda, T., Arai, F., and Dong, L. (2003). Assembly of nanodevices with carbon nanotubes through nanorobotic manipulations. *Proceedings of the IEEE,* 91, 1803–1818.

Guthold, M., Falvo, M., Matthews, W. G., Paulson, S., Mullin, J., Lord, S., Erie, D., Washburn, S., Superfine, R., Brooks, F. P., and Taylor, R. M. (1999a). Investigation and modification of molecular structures with the nanomanipulator. *Journal of Molecular Graphics and Modelling,* 17, 187–197.

Guthold, M., Matthews, G., Negishi, A., Taylor, R. M., Erie, D., Brooks, F. P., and Superfine, R. (1999b). Quantitative manipulation of DNA and viruses with the nanomanipulator scanning force microscope. *Surface and Interface Analysis,* 27, 437–443.

Johnstone, R. W. and Parmaswaran, A. (2004). *An Introduction to Surface-Micromachining (Information Technology: Transmission, Processing and Storage).* Norwell, Springer.

Landau, L. D. and Lifshitz, E. M. (1960). *Mechanics (Course of Theoretical Physics).* Burlington, Pergamon Press.

Laval, J., Thomas, D., and Mazeran, P. (2000). Nanobiotechnology and its role in the development of new analytical devices. *Analyst,* 125, 29–33.

Liu, K. L., Wu, C. C., Huang, Y. J., Peng, H. L., Chang, H. Y., Chang, P., Hsu, L., and Yew, T. R. (2008). Novel microchip for in situ TEM imaging of living organisms and bioreactions in aqueous conditions. *Lab on a Chip,* 8, 1915–1921.

Lu, Y. W. and Kim, C. J. (2006). Microhand for biological applications. *Applied Physics Letters,* 89, 3.

Luo, J. K., Flewitt, A. J., Spearing, S. M., Fleck, N. A., and Milne, W. I. (2005). Comparison of microtweezers based on three lateral thermal actuator configurations. *Journal of Micromechanics and Microengineering,* 15, 1294–1302.

Marie, R., Thaysen, J., Christensen, C. B. V., and Boisen, A. (2005). DNA hybridization detected by cantilever-based sensor with integrated piezoresistive read-out. *Micro Total Analysis Systems 2004,* Vol. 2. Cambridge, Royal Society of Chemistry.

Muller, D. J., Janovjak, H., Lehto, T., Kuerschner, L., and Anderson, K. (2002). Observing structure, function and assembly of single proteins by AFM. *Progress in Biophysics and Molecular Biology,* 79, 1–43.

Muratore, R., Lamanna, J., Szulman, E., Kalisz, A., Lamprecht, M., Simon, M., Yu, Z., Xu, N., and Morrison, B. (2009). Bioeffective ultrasound at very low doses: Reversible manipulation of neuronal cell morphology and function *in vitro.* In Ebbini, E. S. (Ed.). *8th International Symposium on Therapeutic Ultrasound.* Melville, American Institute of Physics.

Mølhave, K., Wich, T., Kortschack, A., and Bøggild, P. (2006). Pick-and-place nanomanipulation using microfabricated grippers. *Nanotechnology,* 17, 2434–2441.

O'Malley, M. K. (2007). Principles of human-machine interfaces and interactions. In Zhang, M. and Nelson, B. (Eds.). *Life Science Automation. Fundamentals and Applications.* Norwood, Artech House.

Oberringer, M., Englisch, A., Heinz, B., Gao, H., Martin, T., and Hartmann, U. (2003). Atomic force microscopy and scanning near-field optical microscopy studies on the characterization of human metaphase chromosomes. *European Biophysics Journal with Biophysics Letters,* 32, 620–627.

Park, D. S. W., Nallani, A. K., Cha, D., Lee, G. S., Kim, M. J., Skidmore, G., Lee, J. B., and Lee, J. S. (2010). A sub-micron metallic electrothermal gripper. *Microsystem Technologies: Micro- and Nanosystems Information Storage and Processing Systems,* 16, 367–373.

Purcell, E. (1977). Life at low Reynolds number. *American Journal of Physics,* 45, 3–11.

Ramser, K. and Hanstorp, D. (2010). Optical manipulation for single-cell studies. *Journal of Biophotonics,* 3, 187–206.

Rubio, W. M., Silva, E. C. N., Bordatchev, E. V., and Zeman, M. J. F. (2009). Topology optimized design, microfabrication and characterization of electro-thermally driven microgripper. *Journal of Intelligent Material Systems and Structures,* 20, 669–681.

Rubio-Sierra, J., Heckl, W., and Stark, R. W. (2005). Nanomanipulation by atomic force microscopy. *Advanced Engineering Materials,* 7, 193–196.

Sahu, B., Taylor, C. R., and Leang, K. K. (2010). Emerging challenges of microactuators for nanoscale positioning, assembly, and manipulation. *Journal of Manufacturing Science and Engineering—Transactions of the ASME,* 132, 16.

Sott, K., Eriksson, E., Petelenz, E., and Goksor, M. (2008). Optical systems for single cell analyses. *Expert Opinion on Drug Discovery,* 3, 1323–1344.

Spegel, C., Heiskanen, A., Pedersen, S., Emneus, J., Ruzgas, T., and Taboryski, R. (2008). Fully automated microchip system for the detection of quantal exocytosis from single and small ensembles of cells. *Lab on a Chip,* 8, 323–329.

Trimmer, W. S. N. (1989). Microrobots and micromechanical systems. *Sensors and Actuators,* 19, 267–287.

Wautelet, M. (2001). Scaling laws in the macro-, micro- and nanoworlds. *European Journal of Physics,* 22, 601–611.

2 Atomic Force Microscopy for Liquid Applications

Zachary James Davis
Department of Micro- and Nanotechnology,
Technical University of Denmark, Lyngby

CONTENTS

2.1 INTRODUCTION

This chapter presents how atomic force microscopy (AFM) can be used for liquid applications. AFM is a specific scanning probe microscopy (SPM) technique that exploits the atomic interactions between a hard probe and a sample to create a three-dimensional image of the sample topology. Furthermore, AFM can also be used to map the electrical and mechanical properties of a surface.

First, the basics of AFM will be explained, covering the main components of a standard AFM and describing the most important imaging techniques in detail. Then a list of the challenges of using AFM in liquids will be listed and described, and the technical solutions that have emerged to solve these will be briefly presented. Finally, several examples of AFM applications will be described, which will give more insight into the technical solutions available and highlight the large potential of the AFM technique in the study of biological systems in their native liquid environment.

2.2 BASICS OF AFM

The main components of AFM are the cantilever probe, piezoelectric scanner stage, and readout as seen in Figure 2.1.

The cantilever probe consists of a cantilever beam and a sharp tip at the end. The sharp tip interacts with the sample, and the forces exerted on the sample either bend the cantilever in static mode or perturb the resonant properties in dynamic mode. In order to vibrate the cantilever for dynamic mode, a piezoelectric actuator is usually mounted on the base. The movement of the cantilever can be measured in a variety of methods. The most common method is using an optical readout as shown in Figure 2.1. In this case, a laser beam is reflected off the backside of the cantilever, and any deflection is registered in the segmented photodiode. For z deflections, a simple duel-segmented photodiode can be used; however, twisting of the cantilever can also be measured by using a multiple-segmented photodiode. Typical deflection sensitivities using the optical technique are on the sub Å range, which is one of the main reasons for using this technique.

The scanner is used to precisely position the sample with respect to the probe tip. The most typical configuration has the sample situated on top of the scanner; however, there are several advantages for keeping the sample static and scanning the probe instead. The high-precision requirements of the scanner are met by using piezoelectric ceramics, such as lead zirconate titanate (PZT). The scanner is capable of scanning the sample or probe in the x-y plane and moving in the z plane in order to maintain a set-point interaction between the tip and sample.

The AFM works in the following manner. First, the course positioning is used to situate the probe tip close to the sample. Then, the scanner is used to approach the tip and sample until a set-point interaction is reached, which will depend on the AFM mode. The set-point interaction is chosen based on the properties of the probe

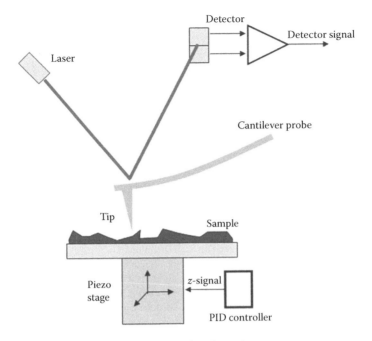

FIGURE 2.1 Typical setup of an AFM, showing the main components.

and sample in order to ensure optimal imaging conditions and to avoid damaging the sample. After approaching the probe and sample, the image is acquired by scanning the sample and monitoring the interactions. If the interactions deviate from the set point, a feedback loop is used to control the scanner in order to maintain the set point. This deviation from the set point is often referred to as the *error signal*. In this way, the probe follows the contour of the sample surface and maintains a constant interaction, which will not damage the probe or sample. The *z* displacement of the scanner is used to generate a 3D image of the sample. The way the probe scans the surface is illustrated in Figure 2.2, where it is seen in the fast-scan direction (*x* axis) and slow-scan direction (*y* axis). Furthermore, the image is taken by using either the trace or retrace, due to the fact that the scanner will always have some nonlinearities.

2.2.1 ATOMIC INTERACTIONS

The atomic interactions between the probe tip and sample can be compared to the interactions between single atoms. The Lennard–Jones potential is a simple mathematical model that describes the interactions between a pair of neutral atoms and is given by the following equation:

$$U(r) \propto \left(\frac{\sigma}{r}\right)^{12} - \left(\frac{\sigma}{r}\right)^{6} \tag{2.1}$$

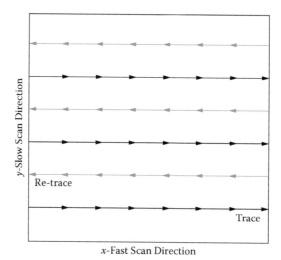

FIGURE 2.2   Illustration of how the AFM scans the sample during imaging, showing the fast-scan direction (*x* axis) and slow-scan direction (*y* axis). The image is taken by using either the trace or retrace due to nonlinearities in the piezoelectric scanner.

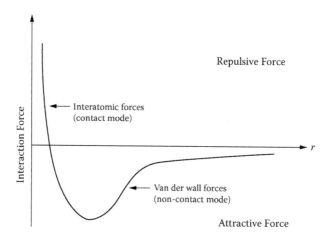

FIGURE 2.3   Force versus atomic distance curve based on the Lennard–Jones potential.

where σ represents the distance at which the potential is zero, and r is the distance between the atoms. It is important to note that this equation is not based on any physical theory but is a mathematical model based on experimental finding.

The force between the atoms is found by differentiating Equation 2.1 with respect to the distance, and the force versus atomic distance curve can be seen in Figure 2.3.

There are two main force regimes:

Van der Waals force—This is the long-distance attractive force, which is caused by the fluctuating dipoles of the atoms.

Interatomic force—This is the short-distance repulsive force, which is due to electrostatic and quantum mechanical interactions when the electron clouds overlap.

2.2.2 Modes of Operation

There are several modes of operation for the AFM. These different modes can also be divided into two groups: static and dynamic. The static mode represents modes where the static deflection of the cantilever is used to measure the probe–sample interaction, whereas the dynamic mode is based on measuring changes in the cantilevers dynamic properties when the cantilever is vibrating close to or at its resonant frequency.

2.2.2.1 Contact AFM

A schematic diagram of the contact AFM is shown in Figure 2.4. In contact mode, the tip is approached to the sample, and the optical readout measures the static deflection of the cantilever. When the deflection reaches set-point deflection, the approach stops and scanning is started. The feedback loop ensures that if the deflection deviates from the set point, the error signal is sent to the position integral derivative (PID) controller, which moves the sample up or down to maintain the set-point deflection.

In contact mode, the probe tip is in full contact with the surface. When the tip is in full contact with the sample, the force regime is the repulsive interatomic forces. The forces that are exerted on the probe tip will deflect the cantilever, which acts like a

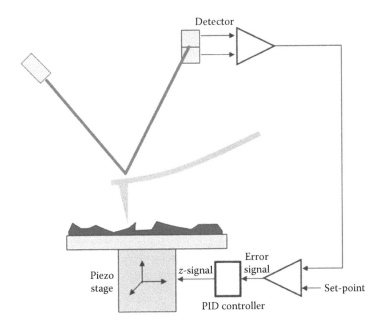

FIGURE 2.4 Schematic diagram of contact AFM. The optical detector measures the static deflection, and the feedback loop compares this value to the set point, which then controls the PID controller to maintain the set point by moving the piezo-stage.

spring. The set-point force can be related to the cantilever deflection (Δz) by knowing the spring constant (k) of the cantilever through Hook's law:.

$$F = k \cdot \Delta z \tag{2.2}$$

The cantilever's spring constant depends on the cantilever's dimensions and material properties and is

$$k = \frac{Ewt^3}{4L^3} \tag{2.3}$$

where E is the Young's modulus of the cantilever material, and w, t, and L are the width, thickness, and length of the cantilever, respectively.

The main advantage of contact AFM is that the force regime has the largest force gradient, which in turn allows an extremely high resolution. However, the repulsive inter-atomic forces are also the largest, which means that there is a potential danger of damaging the tip or sample if the cantilever spring constant is stiffer than the sample or tip. This means that when imaging soft biological systems, very soft cantilevers are needed. Usually, contact mode is not used for imaging biological samples, due to the large vertical and lateral interaction forces, which can push and bend soft biological samples and thus distort the AFM image. In this case, either noncontact or tapping modes are used.

2.2.2.2   Noncontact Mode

Noncontact mode is one of many dynamic modes where the cantilever is vibrating close to its resonant frequency above the surface of the sample. The tip is in noncontact, meaning several nanometers above the surface, and thus the interaction forces between the tip and sample are in the attractive van der Waals regime as shown in Figure 2.3. Opposite to contact mode, a stiff cantilever is used with a typical resonant frequency in the hundreds of kilohertz range in order to have a vibration amplitude in the order of a few nanometers.

Instead of static deflection of the cantilever, noncontact mode measures the dynamic properties of the cantilever in order to measure the interactions between the tip and sample. In amplitude modulation AFM, the vibration amplitude close to its resonant frequency is used. The resonant frequency of the cantilever is proportional to the square root of its effective spring constant over its effective mass as seen in the following equation:

$$f = \frac{1}{2\pi} \sqrt{\frac{k_{eff}}{m_{eff}}} \tag{2.4}$$

When the tip approaches the sample, it will feel a positive force gradient in the noncontact regime as seen in Figure 2.3. This force gradient will change the effective spring constant and thus shift the resonant frequency of the cantilever. The effective spring constant, when a force gradient is acting on the tip, is given in the following expression:

$$k_{eff} = k - \frac{\partial F}{\partial z} \tag{2.5}$$

where k is the spring constant of the cantilever beam without any force gradients acting upon it (Equation 2.3), and $\partial F/\partial z$ is the force gradient.

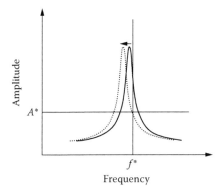

FIGURE 2.5 Frequency response curves illustrating amplitude modulation noncontact AFM.

When the tip–sample separation enters the noncontact regime, which is dominated by attractive van der Waals interactions, the force gradient is positive. This will, according to Equation 2.5, decrease the effective spring constant and thus decrease the resonant frequency of the cantilever. This is illustrated in the frequency responses seen in Figure 2.5. The frequency response shows the amplitude of vibration as a function of the actuation frequency where the peak amplitude corresponds to the cantilever's resonant frequency.

Figures 2.5 and 2.6 illustrate the amplitude modulation (AM) technique in more detail. In AM AFM, first the frequency response is measured when the tip is far away from the sample, and thus no force gradients are present. Then, a set-frequency (f*) is chosen, which is slightly larger than the resonant frequency. Further, a set-amplitude (A*) is chosen, which is lower than the amplitude at the set-frequency. This set amplitude corresponds to a force gradient interaction between the tip and sample under scanning. Upon approaching the tip to the sample, the attractive van der Waals forces will shift the frequency down, which will intern decrease the amplitude at f* until the amplitude reaches the set-amplitude (A*). At this point, the sample is scanned and the feedback loop is used to maintain A* by moving the sample in the z-direction. This z-movement corresponds to the topography of the sample.

By referring again to Figure 2.3, the forces at the noncontact region are much lower than at the contact mode region and the lateral forces are eliminated, and thus the noncontact mode is ideal for imaging soft biological samples. One of the challenges in performing the noncontact mode in liquids, however, is that the liquids will dampen the vibration on the cantilever, which can decrease its Q-factor. This decreases the slope of the frequency response as seen in Figure 2.7, and this in turn decreases the resolution of the AM AFM technique.

2.2.2.3 Tapping Mode

Tapping mode is an alternate dynamic method to noncontact mode. The main difference is that the vibrational amplitude is larger, typically tens of nanometers. In this case, the tip is tapping the surface in full contact with each oscillation of the cantilever.

In tapping mode, the set frequency (f*) is below the resonant frequency instead of above it. In this case, the force gradient, which will decrease the resonant frequency,

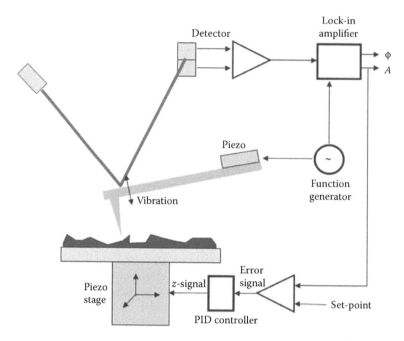

FIGURE 2.6 Schematic diagram of AM AFM typically used for noncontact and tapping mode AFM. The function generator is used to excite the cantilever into vibration, and the lock-in amplifier measures the amplitude and phase at a set-point frequency f*. Then the amplitude is fed back and compared to the set-point value (A*), and the deviation is thus used to control the PID, which then moves the piezo-stage to maintain the set-point amplitude.

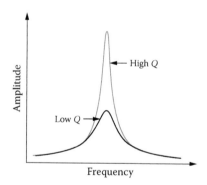

FIGURE 2.7 The frequency response of a cantilever with a high and low Q-factor, illustrating what happens when the cantilever is operated in liquids compared to air.

will increase the amplitude and thus will tap the surface as described earlier. Due to the large vibration amplitude, the force regime stretches from the contact regime to the noncontact regime.

The main advantages of tapping mode AFM (TM AFM) is that the force gradient is much larger compared to noncontact mode AFM, thus making it possible to achieve higher resolution. Furthermore, the lateral forces are reduced compared to contact mode, which also makes it more suitable for imaging biological samples.

2.2.2.4 Frequency Modulation AFM

In 1991, Albrecht and coworkers invented frequency modulation AFM (FM AFM) (Albrecht et al., 1991). In FM AFM, the cantilever is put into an oscillator loop where the vibrational amplitude signal is boosted with gain, phase shifted, and fed back to the actuation as seen in Figure 2.8. The oscillation frequency is determined by the mechanical resonant frequency of the cantilever and is read out from the positive feedback loop using an FM demodulator. The set-point force is then related to a set-point frequency shift that is used to keep the tip–sample interactions constant. In 1995, Giessibl et al. demonstrated atomic resolution of a silicon (111) surface using FM AFM in ultrahigh vacuum (Giessibl, 1995). Atomic resolution was possible by exploiting the high Q-factor of the AFM cantilever in ultrahigh vacuum, which minimizes the frequency noise of the oscillator and thus increases the resolution. Again, the main challenge of using FM AFM in liquids is the large viscous damping of the cantilever, which leads to low Q-factors and thus a higher frequency noise.

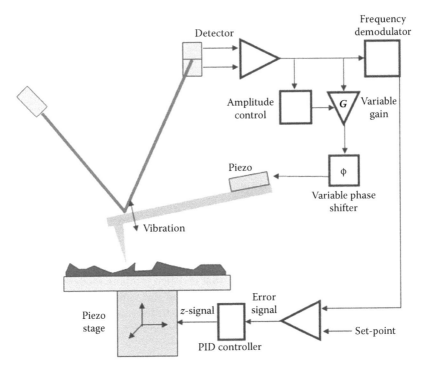

FIGURE 2.8 Schematic diagram of FM AFM, where the cantilever is put into an oscillation loop that oscillates at the resonant frequency of the cantilever. The frequency is measured using an FM demodulator and compared to the set-point frequency. Any deviation from the set point is then sent to the PID that moves the piezo-stage in order to maintain the set-point frequency.

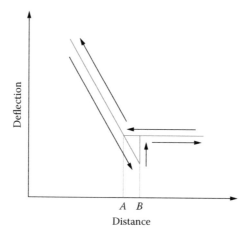

FIGURE 2.9 The deflection versus tip–sample separation distance illustrating the force curve. The deflection is constant until the tip is in contact with the sample at A, then bends with a constant slope. Upon retraction, the adhesion between the tip and sample keeps the tip in contact until point B, whereafter the cantilever snaps to its original position.

2.2.2.5 Force Curves

Another tool of the AFM is force spectroscopy. In this case, the deflection of the AFM cantilever is measured while approaching, contacting, and retracting the probe on the surface at a specific *x,y* location. In this case, the deflection versus distance curves can be used to reveal the adhesion forces between the tip and sample as illustrated in Figure 2.9. As the tip approaches the sample, the deflection is constant until the tip reaches the sample denoted at point A in Figure 2.9. Then the cantilever starts to deflect due to the repulsive interatomic forces. Upon retracting the cantilever from the sample, the deflection decreases; however, due to the adhesion forces between the tip and sample, the tip will stick to the sample until a distance B, where after the force of the cantilever bending the opposite way imposes a larger force than the adhesions force between the tip and sample. By knowing the distance between A and B and knowing the spring constant of the cantilever *k*, the exact adhesion force between the tip and sample can be quantitatively measured.

In 1994, Vanderwerf et al. reported combining this technique while scanning and called it *adhesion mode AFM* (Vanderwerf et al., 1994), and a similar technique was reported in 1998 by de Pablo et al. that is called jumping mode scanning force microscopy, where both topology and adhesion of the sample is mapped (de Pablo et al., 1998). In both cases, the tip is bounced into the sample, and the deflection versus distance curve is used to measure the adhesion forces at each pixel.

2.3 AFM IN LIQUIDS

2.3.1 Challenges of AFM in Liquids

Performing AFM in liquids is not trivial and gives rises to a number of challenges that must be addressed. Furthermore, dynamic mode imaging is more challenging than static mode imaging since vibrating the cantilever into resonance is needed. The major challenges of liquid operation are listed in the following text.

2.3.1.1 Protection of the Scanner

As seen in Figure 2.1, in most AFMs the sample is scanned by a ceramic piezoelectric scanner. Operation in liquids means that the sample is submersed into liquids, and therefore care must be taken in order to prevent liquids getting into the scanner head. This is usually done by using polymer liquid cells that encapsulate the sample, and/or by using drops of water on the sample. In an ideal case, using a microfluidic chip would be preferable in order to be able to control and flow liquids in and out of the sample chamber.

2.3.1.2 Spurious Resonant Modes

The typical mechanical piezoelectric actuation of the AFM cantilever, as shown in Figure 2.6, will also shake the AFM holder and thus will also be transferred to the liquid cell. This will actuate spurious modes in the system that can make it difficult if not impossible to locate the cantilever resonance.

2.3.1.3 Reduced Sensitivity in Dynamic Mode

In dynamic mode, the cantilever will be damped by the liquid surrounding it. This will decrease and broaden the frequency response as seen in Figure 2.7. It will in turn decrease the amplitude slope when operating in AM AFM, which will result in a reduction in sensitivity. In FM AFM, the damping will increase the frequency noise and require a larger gain in order to achieve oscillation, which will also result in reduction in sensitivity.

2.3.1.4 False Deflection Signals

When operating in liquids, changes in the refractive index during operation will bend the laser beam, causing false signals.

2.3.2 Technical Advances for AFM in Liquids

In order to tackle these challenges, several solutions have been presented. The most noted techniques are described in the following text.

2.3.2.1 Magnetic Actuation

In order to reduce spurious modes, magnetic cantilever actuation has been implemented, proven, and commercialized. In normal AFM, dynamic actuation is performed by mounting the AFM probe onto a piezoelectric actuator, which mechanically shakes the AFM cantilever into resonance. Magnetic actuation requires that the AFM cantilever is fabricated using a magnetic material and applying an alternating magnetic field by use of a coil as seen in Figure 2.10. In this manner, only the cantilever is actuated and not the entire cantilever chip and holder.

2.3.2.2 Q-Control

Another method to filter out spurious modes and to improve sensitivity is to enhance the AFM cantilever's Q-factor electrically. In 2001, Tamayo et al. published the first

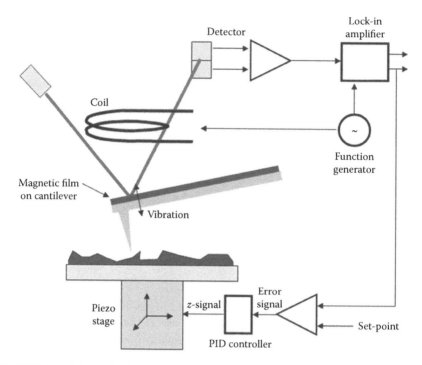

FIGURE 2.10 Schematic illustration of noncontact (NC) AFM using magnetic actuation.

work on the implementation of electrical enhancement of the cantilever Q-factor toward AFM imaging (Tamayo et al., 2001). In "Q-control," the cantilever is put into a positive feedback loop as shown if Figure 2.11a. In this case, the electrical frequency response is enhanced as shown in Figure 2.11b. The AFM is then run in AM AFM mode where the enhanced frequency response is used.

2.3.2.3 High Mechanical Q AFM Probes

Even though Q-control improves AFM by boosting the Q-factor, the signal-to-noise ratio is not improved since both the noise and the signal are amplified through the gain. In order to improve the signal-to-noise ratio, the only method is to increase the Q-factor of the AFM probe itself.

2.3.2.4 Integrated Actuation and Readout

AFM in air and vacuum is performed routinely, and operation in these media is quite straightforward enough for nonexperts to handle. One drawback with AFM in liquids is that this is not a straightforward technique, and therefore it demands a high level of experience. In order to simplify the technique, integrated actuation and readout is needed. Integrated readout would eliminate the need for laser alignment, which is a major difficulty for AFM in liquids. Furthermore, integrated readout would eliminate any false signals due to refractive index changes, as discussed earlier.

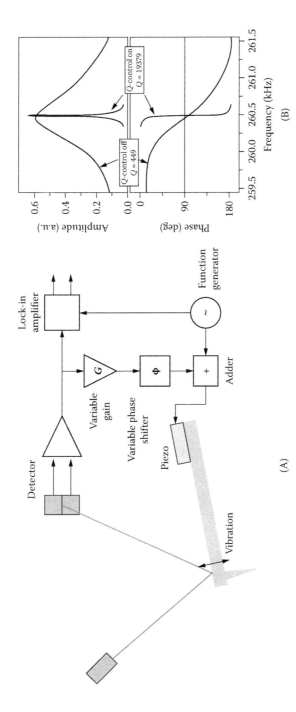

FIGURE 2.11 (A) Schematic diagram of Q-control, which is used to enhance the electrical Q-factor of the cantilever resonance. (B) The magnitude and phase of the cantilever frequency response with and without Q-control showing the Q enhancement. http://www.nanoanalytics.com

2.4 EXAMPLES OF AFM IN LIQUIDS

This section will describe several real examples of researches based on AFM in liquids, illustrating the techniques described in this chapter. A few examples will be described in detail, and other selected examples of the available literature are presented in Table 2.1.

2.4.1 STUDY OF RAT KIDNEY CELLS USING Q-CONTROL AND MAGNETIC ACTUATION

In a paper by Humphris et al. (2000), the use of both Q-control and magnetic actuation was demonstrated on rat kidney cells in liquids. Figure 2.12a and 2.12b compares the frequency spectrums of magnetic and mechanical actuation. In Figure 2.12a, magnetic actuation of the cantilever spectrum shows a single peak corresponding to the cantilever resonance, whereas in Figure 2.12b, normal mechanical actuation, the cantilever signal is buried under spurious modes.

Q-control is also used to enhance the peak of the cantilever response compared to the spurious mode. In Figure 2.12, the frequency response in liquid is measured with and without Q-control, using both mechanical and magnetic actuation methods. When using mechanical actuation, it is seen that the frequency response is dominated by spurious modes (Figure 2.12b). However, by implementing Q-control, the cantilever resonance peak is enhanced and clearly detectable (Figure 2.12d).

Finally, in Figure 2.12a and 2.12c, Q-control is implemented using magnetic actuation, where the response becomes much sharper and can be used to obtain higher-resolution images.

In Figure 2.13, phase shift AFM images of rat kidney cells in water are shown with and without Q-control. Q-control enhances the lateral resolution of the image quite dramatically, proving the merit of this technique.

2.4.2 QUARTZ TUNING FORK AFM PROBES FOR LARGE MECHANICAL Q-FACTORS

Silicon cantilevers, which are normally used for AFM, have typical values of Q on the order of 50–100 in air and can drop down to under 5 in liquids. Hida et al. have reported in 2008 a quartz tuning fork probe with integrated AFM tip (Hida et al., 2008). In air, the tuning fork has achieved Q-factors up to 5000, which is more than an order of magnitude better than silicon cantilever-based probes. The micromachined quartz tuning fork also has integrated actuation since the material is piezoelectric. Some SEM images of the quartz as well as some measured frequency responses are seen in Figure 2.14. The probe is actuated using the electrodes seen in Figure 2.14a, and the readout was performed optically.

2.4.3 IMAGING OF YEAST CELLS USING SELF-ACTUATED AND SELF-SENSING CANTILEVERS IN LIQUIDS

One recent example of a self-actuating and self-sensing cantilever for liquid operation is presented by Fantner et al. (2009). Figure 2.15 shows an SEM image and schematic

TABLE 2.1

Some Representative Examples from the Literature for Manipulation of Biological Particles with AFM

Sample	Medium	Comments	Reference
Viruses and proteins	Air and liquid	Reviews	(Baclayon et al., 2010a, 2010b)
Microbial cells, surface proteins	Air and liquid	Review	(Scheuring and Dufrene, 2010)
DNA and viruses	Liquid	Rupture force of DNA was measured for first time	Guthold 1999b
Bacterial cells	Air and liquid	Bacterial cells were studied in native soil. It was possible to distinguish cells from mineral particles	(Bolshakova et al., 2010)
Microbial cells, biofilms	Air and liquid	Force measurements of microbial biofilm, mechanical properties of microbial cells	(Wright et al., 2010)
Tobacco mosaic virus	Air	Friction and mechanical properties	(Falvo et al., 1997)
Hydrophobic bacteria	Air	Analysis of force interactions	(Dorobantu et al., 2009; Dorobantu and Gray, 2010)
Single proteins	Liquid	Unfolding of single proteins	(Muller et al., 2002)
Neural cells and DNA	Air and liquid	Manipulation of living neuron cells	(Li et al., 2005)
Microbial surfaces	Air and liquid	Reviews	(Dufrene, 2004; Kaminskyj and Dahms, 2008)
Biological material: cells, proteins, receptors	Air and liquid	Review	(Muller and Dufrene, 2008)
Fibrin, virus, and DNA	Liquid	Estimation of Young's modulus and rupture forces	(Guthold et al., 1999a; Guthold et al., 1999b)
Single DNA strands	Liquid	Writing of micrometer-sized patterns	(Hards et al., 2005)
Intracellular stress fibers	Liquid	Measuring of force curves	(Machida et al., 2010)
Human mesenchymal stem cell	Liquid	Gene transfection into stem cells	(Nakamura et al., 2006)
BALB/3T3 cells	Liquid	Intracellular injection	(Nishida et al., 2002)
Mycobacteria	Liquid	Binding strength measurements	(Verbelen and Dufrene, 2009)
Chromosome	Air	Dissection, slide, and pickup of metaphase chromosomes	(Yamanaka et al., 2008)

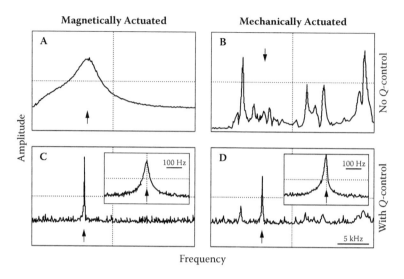

FIGURE 2.12   Frequency responses for a cantilever using magnetic actuation (A and C) and mechanical actuation (B and D), with (A and B) and without (C and D) Q-control. The arrow corresponds to the resonant frequency of the AFM cantilever. (Reprinted from Humphris, A. D. L., Tamayo, J., and Miles, M. J. *Langmuir,* 16, 7891–7894, 2000. With permission. Copyright 2000 American Chemical Society.)

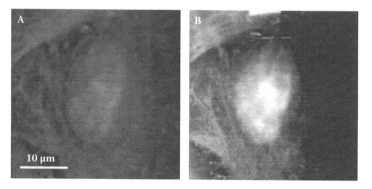

FIGURE 2.13   Tapping-mode phase-shift images of rat kidney cells in liquid without (A) and with (B) Q-control. (Reprinted from J. Tamayo, A. D. L. Humphris, R. J. Owen, M. J. Miles, *Biophysical Journal,* 81, 526–537, 2001. Copyright 2001, with permission from Elsevier.)

diagram of the AFM cantilever. The cantilever is actuated into resonance using a built-in thermal actuator at the free end of the cantilever. By running an AC current through this resistor, the cantilever is heated and bends due to the bimorph effect with the same frequency as the AC signal. When the frequency matches the resonant frequency of the cantilever, it will resonate. At the base of the cantilever, two silicon piezoresistors are placed in order to measure the cantilevers' displacement. The other two resistors on the cantilever base are then used to form a Wheatstone bridge, which is used to convert the resistance change into a voltage signal.

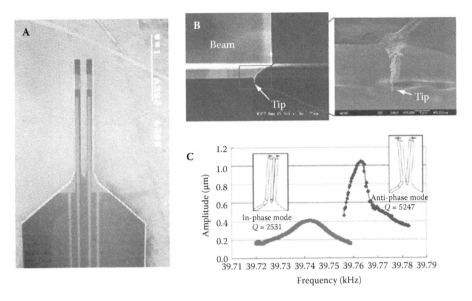

FIGURE 2.14 (A and B) SEM images of a quartz tuning fork–based AFM probe with integrated tip. (C) Frequency responses of the tuning fork in the in and antiphase modes, demonstrating Q-factors on the order of 5000 in air. (Reprinted from Hida, H., Shikida, M., Fukuzawa, K., Murakami, S., Sato, K., Asaumi, K., Iriye, Y., and Sato, K. *Sensors and Actuators a-Physical,* 148, 311–318, 2008. Copyright 2008. With permission from Elsevier.)

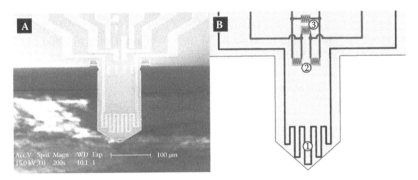

FIGURE 2.15 SEM image (A) and schematic drawing (B) of the AFM cantilever with integrated thermal actuation and piezoresistive readout. (1) Shows the thermal actuator and (2) the piezoresistors, and (3) are on chips resistors in order to form a Wheatstone bridge. (Reproduced with permission from IOP Publishing from Fantner, G. E., Schumann, W., Barbero, R. J., Deutschinger, A., Todorov, V., Gray, D. S., Belcher, A. M., Rangelow, I. W., and Youcef-Toumi, K. *Nanotechnology,* 20, 10, 2009.)

In the paper, the self-actuating and self-sensing cantilever was benchmarked against mechanical actuation and optical readout. In Figure 2.16, the frequency responses show that the thermal actuator functions better than the mechanical since thermal actuation reduces the actuation of spurious modes, just as with magnetic actuation.

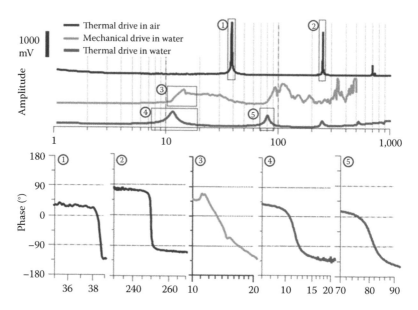

FIGURE 2.16 Frequency responses (magnitude and phase plots) for the AFM cantilever in air and water comparing mechanical drive and thermal drive. (Reproduced with permission from IOP Publishing from Fantner, G. E., Schumann, W., Barbero, R. J., Deutschinger, A., Todorov, V., Gray, D. S., Belcher, A. M., Rangelow, I. W., and Youcef-Toumi, K. *Nanotechnology*, 20, 10, 2009.)

In Figure 2.17, four tapping-mode AFM images of yeast cells in water are given, demonstrating that, in liquids, integrated actuation and readout (Figure 2.17D) can be used and that it gives comparable imaging capabilities as traditional mechanical actuation and optical readout (Figure 2.17A).

2.4.4 IMAGING OF DNA AND FILAMENTS USING JUMPING MODE AFM IN LIQUIDS

In 2003 and 2004, two articles were published by Moreno-Herrero et al., highlighting the use of jumping mode AFM (JM AFM) in liquids. The first article (Moreno-Herrero et al., 2003) describes the use of JM AFM on imaging DNA in liquids. The motivation to perform imaging of DNA in liquids is to be able to image them in their own environment and because it is very difficult to image them in air due to dehydration, tip compression, or other mechanisms that compress the DNA molecule to the surface, making it difficult to resolve the structural details. By using JM AFM with soft cantilevers usually used for contact mode AFM, the authors demonstrated that JM AFM can easily resolve DNA on mica in liquid. Furthermore, JM AFM can be used without damaging the DNA molecules, which would otherwise be very difficult in standard contact- mode AFM due to the large lateral forces involved.

In Figure 2.18, both the topographic and adhesion images are shown of a DNA adsorbed on mica. The height value of the DNA is approximately 2 nm, corresponding

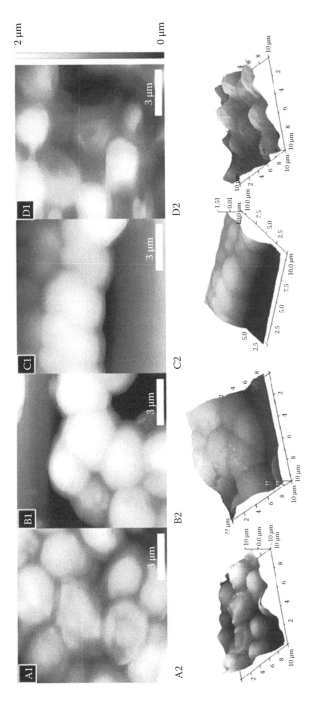

FIGURE 2.17 **(See color insert.)** Tapping-mode AFM images of yeast cells in water using (A) mechanical actuation and optical readout, (B) mechanical actuation and piezoresistive readout, (C) thermal actuation and optical readout, and (D) thermal actuation and piezoresistive readout. (Reproduced with permission from IOP Publishing from Fantner, G. E., Schumann, W., Barbero, R. J., Deutschinger, A., Todorov, V., Gray, D. S., Belcher, A. M., Rangelow, I. W., and Youcef-Toumi, K. *Nanotechnology*, 20, 10, 2009.)

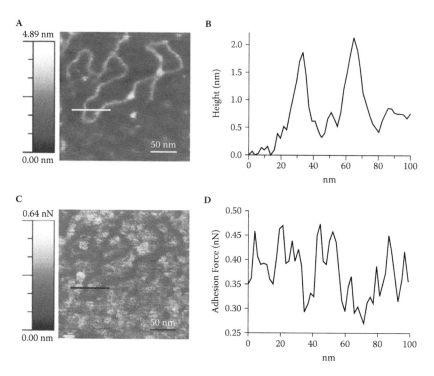

FIGURE 2.18   JM AFM images showing both topographic (A & B) and adhesion (C & D) data. Reprinted from Moreno-Herrero, F., de Pablo, P. J., Alvarez, M., Colchero, J., Gomez-Hertero, J., and Baro, A. M. *Applied Surface Science,* 210, 22–26, 2003. Copyright 2003. With permission from Elsevier.)

nicely with what is expected, and even more interesting, the JM AFM–measured height is larger than dynamic mode AFM (DM AFM)measurements, suggesting that JM AFM is applying less force on the DNA molecule and is therefore more suitable for biological systems in their natural liquid environment.

Another example of JM AFM is the study of Alzheimer paired helical filaments (PHF) in liquids, also published by Moreno-Herrero (2004). In this paper, JM AFM and DM AFM are used, illustrating again that JM AFM provides less interaction forces during imaging, thus imaging the biological structures with less distortions. AFM images of Alzheimer PHF are shown in Figure 2.19. A and B are taken with DM AFM in air, C and D are taken with DM AFM in liquids, and E and F are taken with JM AFM in liquids. First of all, in the DM AFM images in liquids, almost no PHF molecules are present. It is speculated that the molecules are being pushed away during the scanning. However, in the JM AFM image, several PHF molecules can be seen. It is also clearly seen that DM AFM in liquids cannot resolve the helical structure of the single PHF, which is easily seen in DM AFM in air and JM AFM in liquids.

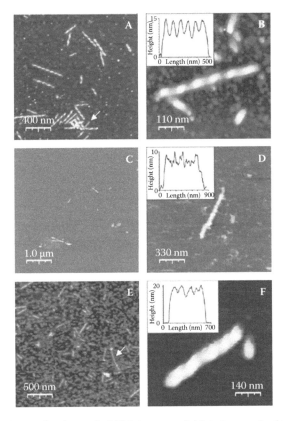

FIGURE 2.19 (See color insert.) AFM images of Alzheimer paired helical filaments using DM AFM in air (A & B), DM AFM in liquid (C & D), and JM AFM in liquid (E & F). Reprinted from Moreno-Herrero, F., Colchero, J., Gomez-Herrero, J., Baro, A. M., and Avila, J. *European Polymer Journal, 40,* 927–932, 2004. Copyright 2004. With permission from Elsevier.)

2.4.5 DIRECT MEASUREMENTS ON MYCOBACTERIUM USING CONTACT AND ADHESION MODE AFM

One major topic of interest is on the study of cell membranes. In the past, both optical and electron microscopy techniques had been utilized. The major disadvantage of optical microscopy is the limited spatial resolution, whereas in electron microscopy the biological samples must be freeze dried and cannot be imaged in their native environment. By using AFM, scientists can now image cells in their natural environment and do so with a very high resolution.

A recent example comes from Verbelen and Dufrene (2009) where they are investigating the cell surface of Mycobacterium, in particular, the fibronectin attachment proteins (FAPs) exposed on the bacterial cell surface. These proteins are responsible for binding to other cell hosts or tissues via fibronectin (Fn). The motivation of this study is to characterize the FAP sites on the bacterial cell membrane and to study

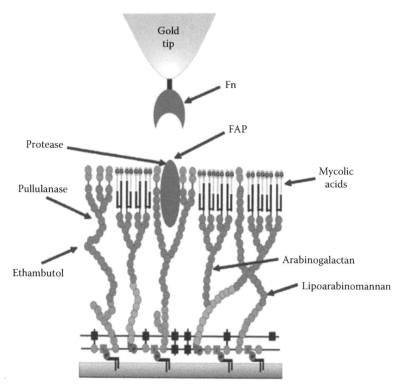

FIGURE 2.20   Schematic drawing of the cell surface including the fibronectin attachment protein (FAP), which is responsible for tissue adhesion through the fibronectin (Fn) molecule. In order to image the FAP protein, the AFM tip was directly functionalized with Fn. (Reproduced from Verbelen, C. and Dufrene, Y. F. *Integrative Biology,* 1, 296–300, 2009. With permission of The Royal Society of Chemistry.)

the effects of antiadhesion drug therapies. A model of the cell surface is shown in Figure 2.20, illustrating the FAP sites embedded in the bacterial cell surface.

In Figure 2.21A, contact mode AFM images of the cell surface are seen, showing a very smooth surface, and thus, using this method, the FAP molecules cannot be seen. In order to find the FAP molecules on the surface, the AFM tip is functionalized with Fn. This is done by first coating the AFM tip with Au, which will readily bind to the sulfur that is contained in the Fn molecule. By using this method, Fn is directly functionalized on the AFM tip without the use of any spacer molecule. Figure 2.21B shows an adhesion map, and Figure 2.21C shows the corresponding adhesion force histogram. By using adhesion mode, it is seen that, in 63% of the force curves, an adhesion event took place with an average adhesion force of approximately 50 pN. These adhesion events are due to binding of the Fn molecule to the FAP protein; thus the results show a high degree of coverage of these binding sites.

In Figure 2.22, a time series of height and adhesion maps is seen after the introduction of pullulanase enzymes, which are known to prevent the adhesion of Mycobateria to other cells and tissues by targeting the cell surface. As seen in the figure, the surface roughness increases from 0.3 nm to 1.8 nm, and the percentage of adhesion

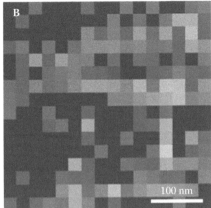
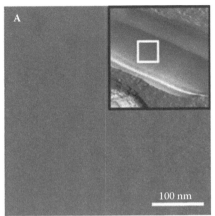
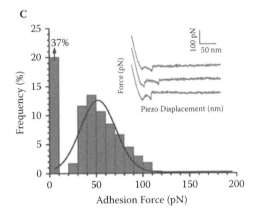

FIGURE 2.21 (A) *In vivo*, contact-mode AFM images of a Mycobacterium surface. (B) Adhesion map of the same surface showing the adhesion events in grayscale, and (C) adhesion force histogram showing a large percentage of adhesion events, each corresponding to an Fn–FAP interaction. (Reproduced from Verbelen, C. and Dufrene, Y. F. *Integrative Biology*, 1, 296–300, 2009. With permission of The Royal Society of Chemistry.)

events decreases dramatically from 80% to 26% after approximately 70 min. These results illustrate the potential of AFM for not only studying the morphology of cell membranes but also drug screening, where dynamic processes can be observed *in vivo* and in real time.

2.4.6 ATOMIC AND MOLECULAR RESOLUTION IN LIQUIDS

As stated earlier, FM AFM has been used to achieve atomic resolution in ultrahigh vacuum conditions due to the high Q-factor of the AFM cantilevers. In liquids, the Q-factor of the cantilever drops dramatically and, consequently, the resolution drops as well. In 2005, however, Fukuma et al. demonstrated, for the first time, atomic resolution in water using the FM AFM technique (Fukuma et al., 2005a, 2005b). By using a deflection sensor with a very low noise, they were able to perform FM AFM

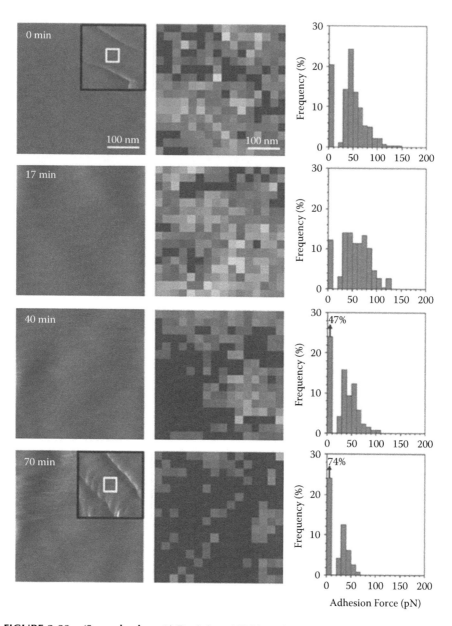

FIGURE 2.22   (See color insert.) Real-time AFM imaging of the Mycobacterium surface after the introduction of pullulanase enzyme, showing the slow change in both surface roughness and decrease in adhesion events. (Reproduced from Verbelen, C. and Dufrene, Y. F. *Integrative Biology,* 1, 296–300, 2009. With permission of The Royal Society of Chemistry.)

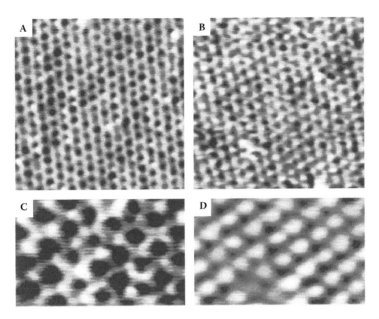

FIGURE 2.23 FM AFM image of cleaved (001) muscovite mica surface in water demonstrating atomic resolution. Reprinted from Fukuma, T., Kobayashi, K., Matsushige, K., and Yamada, H. *Applied Physics Letters,* 87, 3, 2005a. Copyright 2005. With permission from American Institute of Physics.)

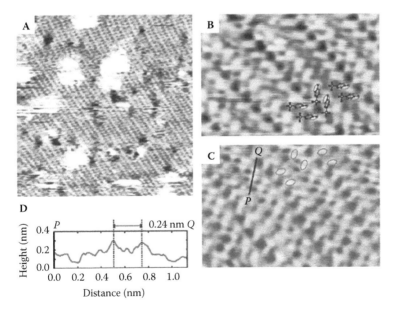

FIGURE 2.24 FM AFM images of polydiacetylene in water, demonstrating molecular resolution on biological samples. (Reprinted from Fukuma, T., Kobayashi, K., Matsushige, K., and Yamada, H. *Applied Physics Letters,* 86, 3, 2005b. Copyright 2005. With permission from American Institute of Physics.)

with extremely small oscillation amplitudes on the order of 0.2–0.3 nm in liquid. The small amplitude enhances the sensitivity to the short-range interaction forces and makes possible atomic resolution in liquids in spite of the fact that the Q-factor of the AFM cantilever is reduced to around 20–30.

In Figure 2.23, images of a cleaved (001) muscovite mica surface are seen. The patterns seen in the images match that of the atomic scale structure of mica and thus demonstrate atomic resolution in liquids.

Fukuma et al. also demonstrated molecular resolution by imaging polydiacety-lene in water using the same FM AFM method as seen in Figure 2.24.

2.5   CONCLUSIONS

In this chapter, the AFM technique and its use for liquid applications have been presented and discussed. Several AFM modes have been described, such as con-tact, noncontact, tapping, and force modulation modes. Furthermore, force spec-troscopy has been introduced, which can be used for adhesion and jumping modes that can be used to measure not only the height but also the adhesion between the tip and sample.

A list of the challenges when using AFM in liquids has been presented along with the technical solutions that have been implemented. One of the major advancements has been the implementation of integrated actuation (magnetic or other), which has rev-olutionized AFM for liquid operation and is now a commercially available product.

Finally, several examples of AFM have been presented in Section 2.3.1, highlighting some of the technical advances and also showing the huge potential for AFM in liquids. Being able to image biological structures in their native environment it has already led to advancements in the understanding of DNA and cell membrane structure and a myriad of other biological mysteries, and will keep doing so in the decades to follow.

REFERENCES

Albrecht, T. R., Grutter, P., Horne, D., and Rugar, D. (1991). Frequency-modulation detection using high-Q cantilevers for enhanced force microscope sensitivity. *Journal of Applied Physics,* 69, 668–673.

Baclayon, M., Roos, W. H., and Wuite, G. J. L. (2010a). Sampling protein form and function with the atomic force microscope. *Molecular and Cellular Proteomics,* 9, 1678–1688.

Baclayon, M., Wuite, G. J. L., and Roos, W. H. (2010b). Imaging and manipulation of single viruses by atomic force microscopy. *Soft Matter,* 6, 5273–5285.

Bolshakova, A. V., Vorobyova, E. A., and Yaminsky, I. V. (2010). Atomic force microscopy studies of living bacterial cells in native soil and permafrost. *Materials Science and Engineering B-Advanced Functional Solid-State Materials,* 169, 33–35.

de Pablo, P. J., Colchero, J., Gomez-Herrero, J., and Baro, A. M. (1998). Jumping mode scan-ning force microscopy. *Applied Physics Letters,* 73, 3300–3302.

Dorobantu, L. S., Bhattacharjee, S., Foght, J. M., and Gray, M. R. (2009). Analysis of force interactions between AFM tips and hydrophobic bacteria using DLVO theory. *Langmuir,* 25, 6968–6976.

Dorobantu, L. S. and Gray, M. R. (2010). Application of atomic force microscopy in bacterial research. *Scanning,* 32, 74–96.

Dufrene, Y. F. (2004). Using nanotechniques to explore microbial surfaces. *Nature Reviews Microbiology,* 2, 451–460.

Falvo, M. R., Washburn, S., Superfine, R., Finch, M., Brooks, F. P., Chi, V., and Taylor, R. M. (1997). Manipulation of individual viruses: Friction and mechanical properties. *Biophysical Journal,* 72, 1396–1403.

Fantner, G. E., Schumann, W., Barbero, R. J., Deutschinger, A., Todorov, V., Gray, D. S., Belcher, A. M., Rangelow, I. W., and Youcef-Toumi, K. (2009). Use of self-actuating and self-sensing cantilevers for imaging biological samples in fluid. *Nanotechnology,* 20, 10.

Fukuma, T., Kobayashi, K., Matsushige, K., and Yamada, H. (2005a). True atomic resolution in liquid by frequency-modulation atomic force microscopy. *Applied Physics Letters,* 87, 3.

Fukuma, T., Kobayashi, K., Matsushige, K., and Yamada, H. (2005b). True molecular resolution in liquid by frequency-modulation atomic force microscopy. *Applied Physics Letters,* 86, 3.

Giessibl, F. J. (1995). Atomic-resolution of the silicon (111)-(7×7) surface by atomic-force microscopy. *Science,* 267, 68–71.

Guthold, M., Falvo, M., Matthews, W. G., Paulson, S., Mullin, J., Lord, S., Erie, D., Washburn, S., Superfine, R., Brooks, F. P., and Taylor, R. M. (1999a). Investigation and modification of molecular structures with the nanomanipulator. *Journal of Molecular Graphics and Modelling,* 17, 187–197.

Guthold, M., Matthews, G., Negishi, A., Taylor, R. M., Erie, D., Brooks, F. P., and Superfine, R. (1999b). Quantitative manipulation of DNA and viruses with the nanomanipulator scanning force microscope. *Surface and Interface Analysis,* 27, 437–443.

Hards, A., Zhou, C. Q., Seitz, M., Brauchle, C., and Zumbusch, A. (2005). Simultaneous AFM manipulation and fluorescence imaging of single DNA strands. *Chemphyschem,* 6, 534–540.

Hida, H., Shikida, M., Fukuzawa, K., Murakami, S., Sato, K., Asaumi, K., Iriye, Y., and Sato, K. (2008). Fabrication of a quartz tuning-fork probe with a sharp tip for AFM systems. *Sensors and Actuators a-Physical,* 148, 311–318.

Humphris, A. D. L., Tamayo, J., and Miles, M. J. (2000). Active quality factor control in liquids for force spectroscopy. *Langmuir,* 16, 7891–7894.

Kaminskyj, S. G. W. and Dahms, T. E. S. (2008). High spatial resolution surface imaging and analysis of fungal cells using SEM and AFM. *Micron,* 39, 349–361.

Li, G., Xi, N., and Wang, D. H. (2005). In situ sensing and manipulation of molecules in biological samples using a nanorobotic system. *Nanomedicine: Nanotechnology, Biology, and Medicine,* 1, 31–40.

Machida, S., Watanabe-Nakayama, T., Harada, I., Afrin, R., Nakayama, T., and Ikai, A. (2010). Direct manipulation of intracellular stress fibres using a hook-shaped AFM probe. *Nanotechnology,* 21, 385102-1–385102-6.

Moreno-Herrero, F., Colchero, J., Gomez-Herrero, J., Baro, A. M., and Avila, J. (2004). Jumping mode atomic force microscopy obtains reproducible images of Alzheimer paired helical filaments in liquids. *European Polymer Journal,* 40, 927–932.

Moreno-Herrero, F., de Pablo, P. J., Alvarez, M., Colchero, J., Gomez-Hertero, J., and Baro, A. M. (2003). Jumping mode scanning force microscopy: a suitable technique for imaging DNA in liquids. *Applied Surface Science,* 210, 22–26.

Muller, D. J. and Dufrene, Y. F. (2008). Atomic force microscopy as a multifunctional molecular toolbox in nanobiotechnology. *Nature Nanotechnology,* 3, 261–269.

Muller, D. J., Janovjak, H., Lehto, T., Kuerschner, L., and Anderson, K. (2002). Observing structure, function and assembly of single proteins by AFM. *Progress in Biophysics and Molecular Biology,* 79, 1–43.

Nakamura, C., Han, S., Obataya, I., Imai, Y., Nagamune, T., and Miyake, J. (2006). A novel living cell manipulation technology using nanoneedle and AFM. In Zhou, G., Lu, Z., and Takeyama, H. (Eds.) 2006. *Progress on Post-Genome Technologies.* Conference: 4th Int. Forum on Post-Genome Technologies, Hangzhou, China.

Nishida, S., Funabashi, Y., and Ikai, A. (2002). Combination of AFM with an objective-type total internal reflection fluorescence microscope (TIRFM) for nanomanipulation of single cells. *Ultramicroscopy,* 91, 269–274.

Scheuring, S. and Dufrene, Y. F. (2010). Atomic force microscopy: Probing the spatial organization, interactions and elasticity of microbial cell envelopes at molecular resolution. *Molecular Microbiology,* 75, 1327–1336.

Tamayo, J., Humphris, A. D. L., Owen, R. J., and Miles, M. J. (2001). High-Q dynamic force microscopy in liquid and its application to living cells. *Biophysical Journal,* 81, 526–537.

Vanderwerf, K. O., Putman, C. A. J., Degrooth, B. G., and Greve, J. (1994). Adhesion force imaging in air and liquid by adhesion mode atomic-force microscopy. *Applied Physics Letters,* 65, 1195–1197.

Verbelen, C. and Dufrene, Y. F. (2009). Direct measurement of Mycobacterium–fibronectin interactions. *Integrative Biology,* 1, 296–300.

Wright, C. J., Shah, M. K., Powell, L. C., and Armstrong, I. (2010). Application of AFM from microbial cell to biofilm. *Scanning,* 32, 134–149.

Yamanaka, K., Saito, M., Shichiri, M., Sugiyama, S., Takamura, Y., Hashiguchi, G., and Tamiya, E. (2008). AFM picking-up manipulation of the metaphase chromosome fragment by using the tweezers-type probe. *Ultramicroscopy,* 108, 847–854.

3 Manipulation by Electrical Fields

Maria Dimaki
Department of Micro- and Nanotechnology,
Technical University of Denmark, Lyngby

CONTENTS

The controlled manipulation of biological samples and the liquids they are dispersed in is a large research field within lab-on-a-chip systems. As society is pushing toward faster, cheaper, and reliable biochemical analysis of patient samples, research has been concentrating on miniaturizing existing processes in small chips, thus reducing analysis time and sample volume. This procedure, however, increases the need for sample handling in terms of mixing, pumping, and concentrating. One of the most efficient methods for this purpose is electrokinetics.

Electrokinetics covers a number of different phenomena occurring when an electric field is applied inside a fluid. Direct current (DC) or alternating current (AC) fields can be used. Due to several problems associated with the use of DC fields, such as the need for high voltages that can create bubbles and start unwanted electrochemical reactions, AC fields have been preferred.

There are mainly three types of phenomena associated with AC fields: dielectrophoresis (DEP), AC electroosmosis, and AC electrothermal effects. DEP is the movement of a polarizable particle in an inhomogeneous electric field, and was discovered in 1958 by Pohl. It is by far the most commonly used method for the manipulation of biological structures, so this chapter will concentrate on this. For completeness, AC electroosmosis is the fluid motion induced by moving charges in the double layer, while AC electrothermal effects refer to fluid motion caused by the interaction of electric fields with conductivity and permittivity gradients in the fluid developed by Joule heating.

In the first part of this chapter, the theory behind DEP will be presented, with the main focus on particles with shapes resembling biological samples such as cells and viruses. The second part will attempt to present the different possibilities for biological sample manipulation with this method through examples from the literature.

3.1 UNDERSTANDING THE CONCEPT

As mentioned earlier, DEP is defined as the movement of a polarizable particle in an inhomogeneous electric field. The various parts of this definition will be examined here.

What do we mean by "polarizable?" Polarizability is a measure of the ability of a material to respond to a field (polarize), but it is also a measure of the ability of a material to produce charge at interfaces. When you have a particle in a liquid, then the result of applying an electric field on the system depends on the charge accumulating at the interface of these two, which in turns depends on the electrical properties of both the particle and the liquid it is in. The properties we are referring to are conductivity and permittivity. Conductivity is a measure of the ease with which charge can move through a material, while permittivity is a measure of the energy storage or charge accumulation (at interfaces) in a system.

When an electric field is applied in a solution, the electrical force will move charges from the liquid and the particle, and they will pile up at either side of the interface between the particle and the liquid. The amount of charge depends on the electrical properties of both the particle and the liquid, as well as the field strength. Let us consider a uniform electric field such as that in Figure 3.1. Three things may happen: (1) The polarizability of the particle is greater than that of the liquid (Figure 3.1a). That means more charges accumulate on the inside of the particle than on the outside and, therefore, there is a surplus of minus charges on one side of the particle and plus charges on the other, forming a dipole aligned with the electric field. (2) The polarizability of the particle is smaller than that of the liquid (Figure 3.1b). Now more charges accumulate on the outside of the particle, and this gives rise to a dipole again but with a different direction. (3) The polarizability of the particle is equal to

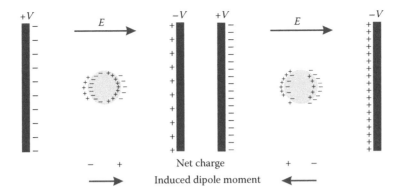

FIGURE 3.1 Dipole formation due to application of a uniform electric field.

that of the liquid (not shown). Here, there are no surplus charges either in or out of the particle, so no dipole is formed.

It should be noted that the charges do not move instantly but take a few microseconds. If you were changing the direction of the field with a certain frequency, then the charges would be able to keep up as long as that frequency was not very high. If you were changing the direction faster than that, though, the charges would not be able to reach equilibrium, and another mechanism would take over for creating interface charges.

The definition of DEP calls for an inhomogeneous (nonuniform) electric field. The field of Figure 3.1 is uniform. This means that the field lines (imaginary lines similar to what you see if you throw iron powder over a magnet) are equally dense on either side of the particle. As the electrical force is proportional to the density of the electric field lines, the force on either side of the particle is the same, so there is no net force pulling the particle anywhere. The particle will not move.

Now consider the case of a nonuniform electric field. Figure 3.2 plots such a field and the resulting field lines for the two cases of a particle being more (Figure 3.2a) or less (Figure 3.2b) polarizable than the liquid. Here, we see that the field lines are denser on one side of the particle than on the other; the electrical force will therefore be stronger on one side of the particle than the other, and the particle will move. In the case of Figure 3.2a the particle will move toward the strongest field (this is called positive DEP), while in the case of Figure 3.2b it will move toward the weakest field (negative DEP).

If the direction of the electric field changes, as it does every half cycle in an AC field, the force on the particle will remain unchanged, as the induced charges will also change sides, as illustrated in Figure 3.3.

3.1.1 GOVERNING EQUATIONS

The force exerted by an electric field $\mathbf{E}$ on a dipole with dipole moment $\mathbf{p}$ is given by

$$\mathbf{F} = (\mathbf{p} \cdot \nabla)\mathbf{E} \tag{3.1}$$

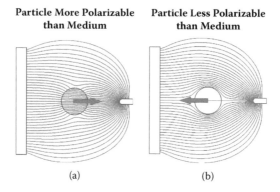

FIGURE 3.2   The field lines in the solution are plotted upon applying a positive voltage on the left electrode and a negative voltage on the right electrode. The field is nonuniform, and a net force appears moving the particle toward (left) or opposite (right) the strongest field gradient.

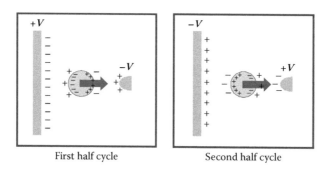

FIGURE 3.3   When the electric field changes direction every half cycle in the AC case, the direction of the induced DEP force remains unchanged.

Here, the higher-order terms of the force are omitted (Morgan and Green, 2003). Thus, the expression is only accurate if the magnitude of the electric field does not vary significantly across the dipole. In an AC field, the time-averaged force on a particle will be given by

$$\mathbf{F}_{\mathrm{DEP}} = \Gamma \cdot \varepsilon_m \cdot \varepsilon_0 \cdot \mathrm{Re}\{K_f\} \cdot \nabla |\mathbf{E}|^2 \tag{3.2}$$

where Γ is a factor depending on the particle geometry, ε_m is the relative permittivity of the suspending medium, ε_0 is the permittivity of free space, and $\mathbf{E}$ is the applied electric field (amplitude and *not* root mean square). The factor K_f depends on the complex permittivities of both the particle and the medium, as well as on the particle geometry. It should be mentioned here that Equation 3.2 is known as the dipole approximation for the dielectrophoretic force, which means that it is only valid if the field does not change much across the particle. If this is not the case, then other equations need to be used. The reader is referred to the work of Morgan and Green (2003) and references therein for more detailed information on the subject.

3.1.1.1 Case of a Spherical Particle

In the case of a spherical particle, the frequency-dependent factor is referred to as the Clausius–Mossotti factor and is given by

$$K_f = \frac{\varepsilon_p^* - \varepsilon_m^*}{\varepsilon_p^* + 2\varepsilon_m^*} \tag{3.3}$$

where the subscripts p and m refer to the particle and the medium, respectively, and the complex permittivity ε^* is given by

$$\varepsilon^* = \varepsilon_0\varepsilon - j\frac{\sigma}{\omega} \tag{3.4}$$

where σ is the conductivity, ε the relative permittivity, and ω the angular frequency of the applied electric field.

The geometry factor Γ is given by

$$\Gamma = \frac{3}{4}V_p \xrightarrow{\;sphere\;} \Gamma = \pi \cdot \alpha^3 \tag{3.5}$$

where α is the sphere radius.

It is worth exploring the factor K_f with respect to frequency a bit more since the dependence of this factor on frequency is important for many applications.

By replacing Equation 3.4 *by* Equation 3.3, we can calculate the real and imaginary parts of K_f. It can easily be shown that

$$\mathrm{Re}\{K_f\} = \frac{\varepsilon_0^2(\varepsilon_p - \varepsilon_m)(\varepsilon_p + 2\varepsilon_m) + \dfrac{(\sigma_p - \sigma_m)(\sigma_p + 2\sigma_m)}{\omega^2}}{\varepsilon_0^2(\varepsilon_p + 2\varepsilon_m)^2 + \dfrac{(\sigma_p + 2\sigma_m)^2}{\omega^2}}$$

$$\mathrm{Im}\{K_f\} = \frac{\varepsilon_0}{\omega}\frac{(\varepsilon_p - \varepsilon_m)(\sigma_p + 2\sigma_m) - (\sigma_p - \sigma_m)(\varepsilon_p + 2\varepsilon_m)}{\varepsilon_0^2(\varepsilon_p + 2\varepsilon_m)^2 + \dfrac{(\sigma_p + 2\sigma_m)^2}{\omega^2}} \tag{3.6}$$

Moreover, the low- and high-frequency limits for the real and imaginary parts of K_f are given by

$$\mathrm{Re}\{K_f\}_{\omega\to0} = \frac{\sigma_p - \sigma_m}{\sigma_p + 2\sigma_m}, \quad \mathrm{Re}\{K_f\}_{\omega\to\infty} = \frac{\varepsilon_p - \varepsilon_m}{\varepsilon_p + 2\varepsilon_m}$$

$$\mathrm{Im}\{K_f\}_{\omega\to0} = \mathrm{Im}\{K_f\}_{\omega\to\infty} = 0 \tag{3.7}$$

This means effectively that, for low frequencies, the magnitude of K_f is controlled by the movement of free charges in the medium and particle due to the electric field, that is, the conductivities of the particle and medium. At high frequencies, these charges are incapable of following the fast-changing electric field, and it is, therefore, the movement of the bound interface charges that controls the magnitude of K_f, that is, the permittivities of the particle and medium.

The imaginary part of K_f reaches a maximum value at a particular frequency ω_{MW}, Often referred to as the Maxwell–Wagner relaxation frequency. It can easily be shown that

$$\omega_{MW} = \frac{\sigma_p + 2\sigma_m}{\varepsilon_0(\varepsilon_p + 2\varepsilon_m)}$$

$$\text{Im}\{K_f(\omega = \omega_{MW})\} = \frac{1}{2}\left[\frac{\varepsilon_p - \varepsilon_m}{\varepsilon_p + 2\varepsilon_m} - \frac{\sigma_p - \sigma_m}{\sigma_p + 2\sigma_m}\right]$$

(3.8)

The relaxation time constant $\tau_{MW} = 1/\omega_{MW}$ is the time required for the interfacial free charge to build up to $1/e$ of its final value (i.e., the time to acquire 63.2% of the total polarization).

In Figure 3.4, the real and imaginary parts of the factor K_f are plotted as a function of frequency for the case of a spherical particle, assuming that $\varepsilon_p > \varepsilon_m$ and $\sigma_p < \sigma_m$.

For spherical particles, the magnitude of the real part is bounded by 1 and $-1/2$, while the magnitude of the imaginary part is bounded by $-3/4$ and $3/4$.

Going back to Equation 3.2, we can see that the dielectrophoretic force depends on the real part of K_f. A positive real part means that the force has direction toward increasing electric fields. This is called *positive DEP*, due to the

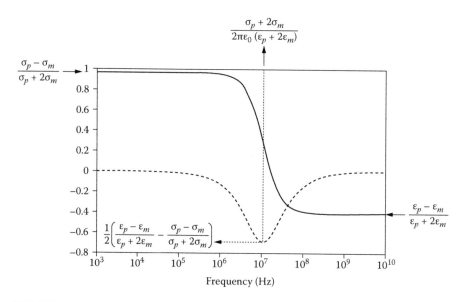

FIGURE 3.4 Plot of the variation of the real (solid line) and imaginary (dotted line) parts of the Clausius–Mossotti factor with frequency. The high- and low-frequency limiting values of the real part are shown, as well as the value of the imaginary part at the relaxation frequency.

positive sign of K_f. If the real part of K_f is negative, then the force on the particle points toward decreasing values of the electric field. The movement is then called *negative DEP*. The frequency at which the behavior changes from positive to negative is called the *turnover frequency* f_{to} and is given by

$$f_{to} = \frac{1}{2\pi \cdot \varepsilon_0} \sqrt{-\frac{\sigma_p - \sigma_m}{\varepsilon_p - \varepsilon_m} \cdot \frac{\sigma_p + 2\sigma_m}{\varepsilon_p + 2\varepsilon_m}}, \quad \begin{cases} \sigma_p < \sigma_m \text{ and } \varepsilon_p > \varepsilon_m \\ \sigma_p > \sigma_m \text{ and } \varepsilon_p < \varepsilon_m \end{cases} \tag{3.9}$$

3.1.1.2 Case of a Prolate Ellipsoid

The simplest case to consider is an ellipsoid, similar to the one shown in Figure 3.5.

The frequency-dependent factor K_f is now different for each principal axis n ($n = 1$, 2 and 3) and is given by

$$K_f^n = \frac{\varepsilon_p^* - \varepsilon_m^*}{\varepsilon_m^* + A_n \cdot \left(\varepsilon_p^* - \varepsilon_m^*\right)} \tag{3.10}$$

A_n is called the depolarizing factor for the axis n, and it is given by

$$A_n = \frac{1}{2}\alpha_1\alpha_2\alpha_3 \int\limits_0^\infty \frac{ds}{\left(s + \alpha_n^2\right) \cdot \sqrt{\left(s + \alpha_1^2\right)\left(s + \alpha_2^2\right)\left(s + \alpha_3^2\right)}} \tag{3.11}$$

where s is an arbitrary distance for integration. The sum of the depolarizing factors along the three principal axes is always unity (Gimsa, 2001) (i.e., $A_1 + A_2 + A_3 = 1$). Each axis of the particle also has an associated relaxation time, which is given by

$$\tau_n = \frac{A_n\varepsilon_p + \left(1 - A_n\right) \cdot \varepsilon_m}{A_n\sigma_p + \left(1 - A_n\right) \cdot \sigma_m} \tag{3.12}$$

The dipole moment of the particle, therefore, depends on its orientation with respect to the electric field. Due to a simultaneously developed torque acting on the particle, one of the principal axes will always orient with the field, but which one it will be depends

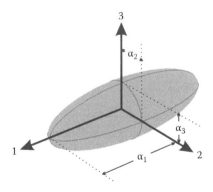

FIGURE 3.5 An ellipsoid with principal axes 1, 2, and 3 and half-lengths along the axes of α_1, α_2, and α_3.

on the frequency of the applied electric field. As a rule of thumb, at a certain frequency the ellipsoid will align with the axis that has the smallest K_f as given by Equation 3.10. Further theoretical analysis of this effect is not relevant to this book, but the interested reader is encouraged to read the attached literature.

The integral of Equation 3.11 is not easy to calculate even with numerical methods. A special (and easier) case to consider is the case of the prolate ellipsoid, where $\alpha_1 \gg \alpha_2 = \alpha_3$. Assuming that the major axis of the particle (axis 1 in this case) is aligned with the field, then the depolarizing factor is given by the simple relationship

$$A_1 = -\frac{1-e^2}{2e^3}\left[2e - \ln\left(\frac{1+e}{1-e}\right)\right]$$

(3.13)

where e is the eccentricity of the particle given by

$$e = \sqrt{1 - \left(\frac{\alpha_2}{\alpha_1}\right)^2}$$

(3.14)

A very useful approximation arises for particles that are really long compared to their thickness. Assuming they can be treated as prolate ellipsoids and when $\alpha_2/\alpha_1 \gg 1$ (typically smaller than 0.001), then from Equation 3.14, $e \approx 1$, and, from Equation 3.13, $A_1 \approx 0$.

Replacing A1 in Equation 3.12, we get that $\tau \approx \varepsilon_m/\sigma_m$. Further, the frequency-dependent factor now becomes

$$K_f^n \approx \frac{\varepsilon_p^* - \varepsilon_m^*}{\varepsilon_m^*}$$

(3.15)

Equation 3.15 is often used instead of Equation 3.10 simply because it is much easier. However, the approximation is not always valid even if $A_1 \approx 0$. Expanding the complex numbers in Equation 3.10, one can show that the approximation is only valid if the conductivity of the particle is not very large. In other words, these relations are only valid if $A_1 \cdot \sigma_p \ll 1$. The error of the approximation gets bigger as the ratio α_2/α_1 gets larger.

The turnover frequency for a prolate ellipsoid can be calculated from Equation 3.10 as

$$f_{to} = \frac{1}{2\pi\varepsilon_0}\sqrt{-\frac{\sigma_p - \sigma_m}{\varepsilon_p - \varepsilon_m}\cdot\frac{\sigma_p + \alpha\sigma_m}{\varepsilon_p + \alpha\varepsilon_m}}, \quad \alpha = \frac{1}{A_n} - 1$$

(3.16)

One can illustrate the errors that can occur by using the approximation of Equation 3.15 by plotting the turnover frequency from Equation 3.16 relative to that given by the approximation as a function of the ratio for various particle conductivities. This is shown in Figure 3.6, where it can be seen that the approximated turnover frequency can be many times smaller than the one that is calculated by the original equation.

In terms of the DEP force on the ellipsoidal particle, Equation 3.2 applies here, too, but the geometry factor Γ is now given by

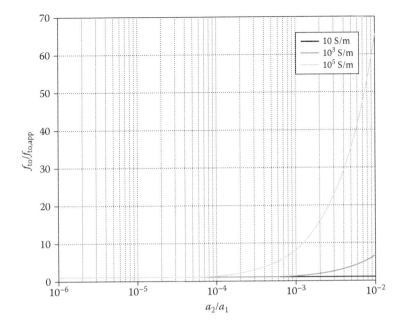

FIGURE 3.6 The turnover frequency compared to the approximated turnover frequency for a particle modeled as a prolate ellipsoid with a permittivity of 60 at a medium with permittivity of 80 and conductivity of 10 mS/m for various values of the ratio and various particle conductivities.

$$\Gamma = \frac{3}{4}V_p \xrightarrow{\ ellipsoid\ } \Gamma = \pi \cdot \alpha_1 \cdot \alpha_2 \cdot \alpha_3 \tag{3.17}$$

The frequency-dependent factor for prolate ellipsoid particles does not show an upper limit of +1, as was the case with spherical particles. The factor can in fact become really large, resulting in large dielectrophoretic forces on this type of particles. An example of a frequency-dependent factor as a function of frequency is shown in Figure 3.7. Here, an ellipsoidal particle with a length of 200 nm and a radius of 10 nm is considered, resulting in an eccentricity of 0.995.

3.1.1.3   Case of a Multilayered Particle

When treating a biological cell under DEP theoretically, one is tempted to view it as a sphere with a certain permittivity and conductivity. This approach is, however, rather simplified. Even the simplest of cells consists of multiple layers: (a) the cell membrane, (b) the cytoplasm, and (c) the nucleus. In some cases, there is even a cell wall around the cell membrane, and some of the organelles inside the cytoplasm can also be treated as a layer. Each layer has its own permittivity and conductivity, but the multishelled particle can be replaced by a homogeneous particle with an effective complex permittivity that relates to those of each shell of the original particle. This effective permittivity can be used in the equations presented in Section 3.1.1.1. Here, the simplest case for an one-shell spherical particle will be presented.

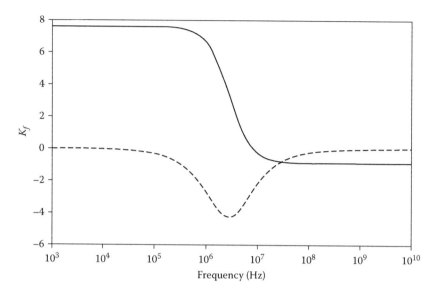

FIGURE 3.7 The real (solid line) and imaginary (dotted line) part of the frequency-dependent factor for a prolate ellipsoid with a radius of 10 nm, a length of 200 nm, permittivity of 10, and conductivity of 0.1 S/m immersed in a water solution of permittivity 78.4 and conductivity of 10 mS/m.

Consider a homogeneous particle with a radius r with a membrane of thickness d surrounding it, as shown in Figure 3.8.

If ε^*_m, ε^*_{int}, and ε^*_{mem} are the complex permittivities of the medium, the inside of the particle, and the membrane around the particle, respectively, then the effective complex permittivity of the particle is given by

$$\varepsilon^*_{eff} = \varepsilon^*_{mem} \cdot \frac{\left(\dfrac{r+d}{r}\right)^3 + 2 \cdot \dfrac{\varepsilon^*_{int} - \varepsilon^*_{mem}}{\varepsilon^*_{int} + 2 \cdot \varepsilon^*_{mem}}}{\left(\dfrac{r+d}{r}\right)^3 - \dfrac{\varepsilon^*_{int} - \varepsilon^*_{mem}}{\varepsilon^*_{int} + 2 \cdot \varepsilon^*_{mem}}} \tag{3.18}$$

The complex permittivities are related to the relative permittivities and conductivities by Equation 3.4.

As an example, Figure 3.9 plots the real and imaginary parts of the frequency-dependent factor for a T-lymphocyte modeled as an one-shell particle.

Characteristic for this type of particles is that they often exhibit more than one turnover frequency.

3.2   PARTICLE SORTING

It is clear from the preceding text that two particles with different electrical properties dispersed in the same liquid would experience a different dielectrophoretic force (DEP force) at the same frequency and could thus be somehow separated. Indeed, separation of particles with different properties is one of the greatest applications of

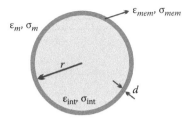

FIGURE 3.8 Schematic of a spherical particle with a single shell.

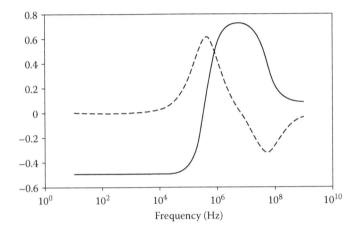

FIGURE 3.9 The real and imaginary part of the effective Clausius–Mossotti factor for a T-lymphocyte modeled as a one-shell particle.

DEP. Knowledge of the particles' permittivity and conductivity, as well as knowledge of their geometry, allows the calculation of the frequency-dependent factor K_f, which in turns gives information on the frequency bands where particles experience positive or negative DEP.

Some biological particles such as various types of white blood cells have strikingly similar sizes and dielectric properties, making separation based on the turnover frequency difficult, if not impossible. In these cases, several techniques have been developed to achieve separation by DEP. Two of them will be presented in the following section.

3.2.1 SORTING OF PARTICLES WITH LARGELY DIFFERENT DIELECTRIC PROPERTIES

In this case, sorting is relatively straightforward once the frequency-dependent factor has been calculated. This can better be explained by looking at Figure 3.10 where the frequency-dependent factor has been plotted for two different particles. The area marked "sorting region" in Figure 3.10 refers to the frequency region, where one of the particles is experiencing positive DEP, while the other experiences negative DEP. Choosing the frequency of the AC electric field in this region would mean that you could separate the two particles since one of them would go toward regions of strong electric field (e.g., the electrodes creating the field) and the other away from such regions.

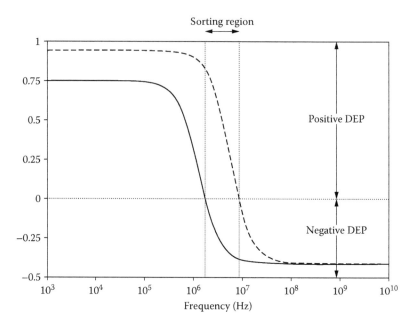

FIGURE 3.10 The Clausius–Mossotti factor plotted against frequency for two different particles. The two particles show distinct differences in their response to an electric field, enabling us to separate them from each other by using the right frequency.

A very interesting parameter to consider when sorting between particles is the turnover frequency. As can be seen from Equations 3.9 and 3.16, the turnover frequency, when it exists, depends on the properties of the medium. Assuming that a particle is capable of existing in solutions of different permittivities and conductivities, we are able to tune the turnover frequency to suit our purpose by changing the properties of the medium. This is illustrated in Figure 3.11 where the turnover frequency as a function of medium conductivity is plotted for a prolate ellipsoid with a relative permittivity of 60 and a conductivity of 0.1 S/m. The medium permittivity is taken to be equal to 80, and its conductivity varies from 1 μS/m to 10 S/m.

As can be seen in the figure, for all practical purposes*, the turnover frequency varies from 3.7 MHz to about 22.5 MHz, depending on medium conductivity. The figure also illustrates that the turnover frequency does not exist for medium conductivities larger than that of the particle.

This figure also illustrates the approximation problem described in Section 3.1.1.2 from the perspective of medium conductivity. When the approximation of Equation 3.15 is used, errors start to appear as the medium conductivity becomes smaller. These errors are bigger as the depolarization factor gets larger, as can be seen by comparing Figure 3.11a and Figure 3.11b. One should, therefore, always use the approximation with care, as it can result in large errors under some circumstances.

---

* In principle, the closest the medium conductivity gets to the conductivity of the particle, the smallest the turnover frequency. The theoretical limit is of course 0 for $\sigma_m = \sigma_p$, but that has no physical meaning. However, when the medium conductivity is only 0.7% smaller than the particle conductivity, the turnover frequency is 3.7 MHz.

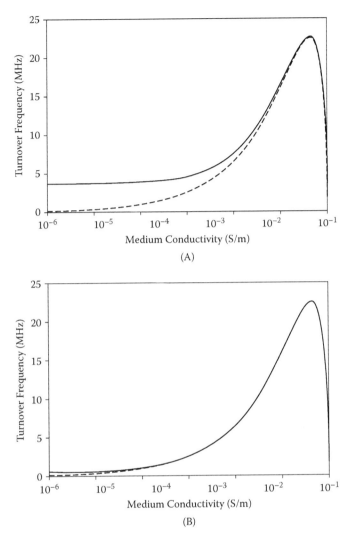

FIGURE 3.11 The turnover frequency as a function of solution conductivity for the case of (A) and (B). The solid line represents the turnover frequency calculated by Equation 3.16, while the dotted line is the turnover frequency calculated using the approximation of Equation 3.15.

3.2.2 MULTIPLE FREQUENCY DIELECTROPHORESIS

Considering the natural biological variation in terms of particle size and properties, the calculated values for the turnover frequency are not absolute but can vary significantly from particle to particle around the theoretical value. This means that particles with small differences in their theoretical turnover frequencies cannot be expected to be sorted just by choosing a frequency between their turnover frequencies.

For such cases, another method has been developed where two or more signals at different frequencies are used to create the electric field. This method has been termed *multiple frequency dielectrophoresis* and, with proper control of the frequencies and amplitudes of the various AC signals, one can "pull" the turnover frequencies further apart so that separation would be possible even taking into account the natural biological variation.

It can be shown (Urdaneta and Smela, 2007) that an effective frequency-dependent factor can be defined for the particles in this case. This factor is now not only dependent on frequency but also on the relative amplitudes of the AC signals as well as on the position of the particles in the system. For n superimposed signals, the effective frequency-dependent factor is given as

$$K_{f,eff} = \frac{\left| \sum_{i=1}^{n} K_{f,i} \cdot \nabla |E_i|^2 \right|}{\left| \sum_{i=1}^{n} \nabla |E_i|^2 \right|} \tag{3.19}$$

where $K_{f,i}$ is the frequency-dependent factor for the particle at the frequency f_i. In the simplest case, the field is composed from two frequencies acting on a single pair of electrodes. In that case, the variation of the field is the same for the two fields, so Equation 3.19 can be simplified to

$$K_{f,eff} = \frac{K_{f,1} \cdot |E_1|^2 + K_{f,2} \cdot |E_2|^2}{|E_1|^2 + |E_2|^2} \Rightarrow K_{f,eff} = \frac{K_{f,1} + K_{f,2} \cdot \kappa^2}{1 + \cdot \kappa^2}, \kappa = \frac{|E_2|}{|E_1|} \tag{3.20}$$

A case where this method can have beneficial effects is the separation of T- and B-lymphocytes. These two types of white blood cells have strikingly similar frequency-dependent factors, as can be seen in Figure 3.12. Indeed, the theoretical difference in their turnover frequencies is of the order of 14 kHz (indicated by "a" in Figure 3.12), which is far too small to have any practical application due to natural biological variation. However, if a signal of 5 kHz at $\kappa = 0.6$ is superimposed, then a larger window appears where separation is possible. This is indicated by "b" in Figure 3.12, which predicts that the two types of cells can now be separated at frequencies larger than 140 MHz.

3.2.3 DEP/Gravitational Field Flow Fractionation

Another technique able to distinguish between particles of similar dielectric properties is the so-called dielectrophoretic/gravitational field flow fractionation (DEP-FFF). This method utilizes the fact that large biological particles, such as cells, sediment when they are placed in fluids as the gravitational force is quite large.

A system such as the one shown in Figure 3.13 is used. The electrodes at the bottom of the channel generate the electric field and, by choosing the right frequency, the DEP force on the particle will be pointing upward, that is, away from regions of strong electric field. The gradient of the electric field in this electrode configuration can be calculated analytically to be (Morgan and Green, 2003)

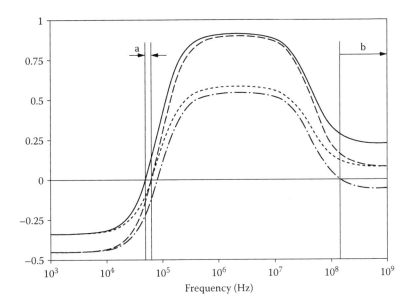

FIGURE 3.12 A plot of the frequency-dependent factor for B-lymphocytes (—) and T-lymphocytes (- - -). When a 5 kHz signal at relative amplitude of 0.6 is superimposed, these plots change—B-lymphocytes (····) and T-lymphocytes (-··-)— and reveal a new region where separation is possible. The dielectric properties of T- and B-lymphocytes are taken from (Yang, J., Huang, Y., Wang, X. J., Wang, X. B., Becker, F. F., and Gascoyne, P. R. C. *Biophysical Journal*, 76, 3307–3314, 1999b). The medium conductivity is set to 10 mS/m, and the medium relative permittivity is set to 80.

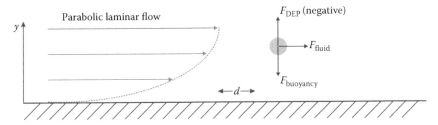

FIGURE 3.13 The forces acting on a particle in a microfluidic channel with electrodes generating an electric field.

$$\nabla |E|^2 = \frac{32}{\pi} \cdot \frac{V_0^2}{d^3} \cdot \exp\left(-\frac{\pi \cdot y}{d}\right) \qquad (3.21)$$

where y is the height of the particle above the electrode plane, d is the width and distance between the electrodes, and $V0$ is the amplitude of the voltage applied to the electrodes. Therefore, the DEP force acting on the particle is given by

$$\mathbf{F_{DEP}} = \pi \alpha^3 \cdot \varepsilon_0 \cdot \varepsilon_m \cdot \mathrm{Re}(K_f) \cdot \nabla |\mathbf{E}|^2 \qquad (3.22)$$

The direction of the buoyancy force depends on the density of the particle compared to that of the fluid. The force is given by

$$\mathbf{F_{buoyancy}} = \frac{4}{3}\pi\alpha^3\left(\rho_p - \rho_m\right)\cdot\mathbf{g} \tag{3.23}$$

where ρ_p is the density of the particle, ρ_m is the density of the medium, and α is the radius of the particle.

Now, depending on the biological particle's properties and the frequency chosen, there will be a height above the electrode plane at which the DEP force and the buoyancy force will be equal. The particle will therefore always remain at that height above the electrode plane as long as it is experiencing both of these forces. The equilibrium point y can be found by equating the magnitudes of the DEP and buoyancy forces as

$$\left|F_{DEP}\right| = \left|F_{buoyancy}\right| \Rightarrow y = \frac{d}{\pi}\ln\left(\frac{24V_0^2\varepsilon_0\varepsilon_m\left|\mathrm{Re}\{CM\}\right|}{\pi d^3\left(\rho_p - \rho_m\right)\cdot g}\right) \tag{3.24}$$

where we have substituted the electric field gradient into the DEP force from Equation 3.21.

Different particles will then balance at different heights above the electrode plane. Since the particles are inside a microfluidic channel operating in a laminar flow regime, the velocity of the liquid at the various particle equilibrium positions, will also be different. Therefore, the particles will flow with different velocities across the channel and will arrive at the outlet at different times, as schematically shown in Figure 3.14.

Particles with very similar dielectric properties are the various types of white blood cells. T- and B-lymphocytes, granulocytes, and monocytes all have turnover

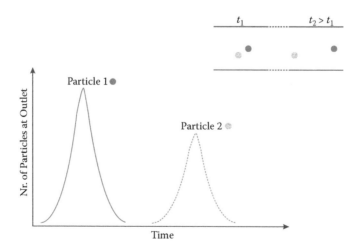

FIGURE 3.14 Schematic of the expected result of an experiment, where particles 1 and 2 are separated using DEP-FFF. The inset shows how the particles will be positioned in the channel at times t_1 and $t_2 > t_1$.

frequencies in the range of 170–350 kHz at a medium conductivity of 56 mS/m (calculated from data found in [Yang et al., 1999b] and references therein). This range is too small to be significant, considering the natural biological variation and the uncertainties associated with the theoretical calculations. However, by use of DEP-FFF, these particles can be separated within 30 min (Yang et al., 1999b; Yang et al., 2000b).

3.3 MOTION OF PARTICLES

Now that the DEP force acting on a particle has been described, it is only natural to continue by describing how the particles will move in a liquid under this force.

In the simplest case, we only consider two forces acting on the particle: the DEP force and the viscous force from the liquid. From Newton's second law of motion, we can then write

$$m\frac{d\mathbf{v}}{dt} = \mathbf{F_{DEP}} + f\cdot(\mathbf{u}-\mathbf{v}) \Rightarrow \mathbf{v} = \left(\frac{\mathbf{F_{DEP}}}{f}+\mathbf{u}\right)\left(1-e^{-\left(f/m\right)t}\right) \qquad (3.25)$$

where m is the particle mass, f is the particle's friction coefficient, and $\mathbf{u}$ is the velocity of the liquid where the particle is moving. The exponential term describes the acceleration term of the particle and has a characteristic time scale of $\tau_a = m/f$. For times greater than τ_a, Equation 3.25 is simplified by discarding the exponential term, and the particle will, therefore, be moving with the terminal velocity given by

$$\mathbf{v} = \frac{\mathbf{F_{DEP}}}{f}+\mathbf{u} \qquad (3.26)$$

Table 3.1 sums up a few of the friction factors for the various cases described earlier. In the table, η is the viscosity of the liquid.

For a more realistic theoretical model, there are a few more considerations. In that case, one can simply replace F_{DEP} in Equation 3.26 by the sum of all forces acting on the particle.

3.3.1 GRAVITATIONAL FORCES

Gravitational forces are usually small but, in some cases, they cannot be ignored, as for example, when cells are under consideration. In fact, one of the most popular cell sorting methods that combines DEP and microfluidics is based on the gravitational forces on cells. This method, called DEP-FFF, was described in detail in Section 3.2.3 for a spherical particle. Generally, the gravitational force (or buoyancy force, as it is otherwise known) on a particle is given by

$$\mathbf{F_{buoyancy}} = V_p \cdot (\rho_p - \rho_m) \cdot \mathbf{g} \qquad (3.27)$$

Here, V_p is the volume of the particle, ρ_p is the density of the particle, ρ_m is the density of the medium, and $\mathbf{g}$ is the acceleration due to gravity.

TABLE 3.1

The Friction Coefficients for a Few Cases of Particles Encountered in the Previous Sections

Particles	Friction Coefficient
Sphere of radius α	$6\pi\eta\alpha$
Prolate ellipsoid moving parallel to the major axis α_1	$\dfrac{8\pi\eta\alpha_1}{2\ln\left(\dfrac{2\alpha_1}{\alpha_2}\right)-1}$
Prolate ellipsoid moving perpendicular to the major axis α_1	$\dfrac{16\pi\eta\alpha_1}{2\ln\left(\dfrac{2\alpha_1}{\alpha_2}\right)-1}$
Prolate ellipsoid moving randomly	$\dfrac{6\pi\eta\alpha_1}{\ln\left(\dfrac{2\alpha_1}{\alpha_2}\right)}$

Source: Modified from Morgan, H. and Green, N. G. (2003). *AC Electrokinetics: Colloids and Nanoparticles*, Research Studies Press. Baldock, United Kingdom.

3.3.2 BROWNIAN MOTION

All particles move randomly in space due to thermal motion, something that is known as *Brownian motion*. Characteristic of this type of movement is the average displacement of the particle over a large number of steps being 0. However, the root mean square of the displacement, Δ, is not 0 but has a value that depends on the diffusion coefficient of the particle and the time of observation. At a given temperature T, the diffusion coefficient of the particle will be given as

$$D = \frac{k_B T}{f} \tag{3.28}$$

where k_B is the Boltzmann constant, f is the friction coefficient, and T is the absolute temperature in Kelvin. For a particle with a diffusion coefficient D that is diffusing in one dimension in time t, the root mean square of the displacement is given as

$$\Delta = \sqrt{2Dt} \tag{3.29}$$

It is easy to deduct that, for a particle moving in three-dimensional space, the root mean square of the displacement is then

$$\Delta = \sqrt{6Dt} \tag{3.30}$$

The effects of the Brownian motion for a particle otherwise moving due to a deterministic force as the DEP force can be understood by introducing the concept of the observably deterministic threshold force (Morgan and Green, 2003). This is the force required to be acting on a particle in order for the observer to be able to tell that there is actually a force acting on the particle.

Consider movement in one dimension. If the root mean square of the displacement for the Brownian motion is given by Equation 3.29, then one can define the observably deterministic threshold force F_{thr} as the force required to move the particle by three times that displacement, so by 3Δ over a time δt. From Equation 3.26, the velocity of the particle under this force would be $v_T = F_{thr}/f$, so after time δt we would get

$$v_T \cdot \delta t = 3\Delta \Rightarrow \frac{F_{thr}}{f} \cdot \delta t = 3 \cdot \sqrt{2D\delta t} \xrightarrow{(24)} F_{thr} = 3 \cdot \sqrt{\frac{2k_B T \cdot f}{\delta t}} \qquad (3.31)$$

For a biological cell at 37°C observed every 10 s, this force is equal to 21.8 fN. This force can be produced by applying 5–10 V (amplitude) on electrodes around 1 mm apart. Placing the electrodes closer together would mean that less voltage would be required.

3.3.3 OTHER FORCES

In the previous discussion, only the forces acting on the particles have been described. As mentioned in the introduction to this chapter, when applying a voltage inside a liquid, forces on the liquid itself will arise. These will inevitably affect the motion of the particles inside the liquid and need to be taken into account when theoretical calculations are being undertaken.

Consider a pair of microelectrodes generating an electrical field with amplitude E. The power generation per unit volume is given by

$$W = \sigma_m E^2 \qquad (3.32)$$

where σ_m is the conductivity of the medium. It follows that the larger the conductivity, the higher the power dissipation and the higher the temperature increase in the liquid containing the particles. This can easily become a problem, especially in the case of biological particles, which are often contained in relatively high conductivity liquids. For a liquid of conductivity 60 mS/m, which is not uncommon for a cell containing liquids in a field of moderate strength around 1 MV/m, the power generation per unit volume will be 6.10^{10} W/m^3, which amounts to 120 W being dissipated in a volume of 2 μL. That is bound to raise the temperature of the liquid with possible fatal consequences for the biological material contained in it.

An order-of-magnitude estimate of the temperature increases due to a voltage Vrms applied between microelectrodes in a liquid of thermal conductivity k, and conductivity σ_m is given by Ramos et al. (1998)

$$\Delta T = \frac{\sigma_m \cdot V_{rms}^2}{k} \qquad (3.33)$$

Applying Equation 3.33 to phosphate-buffered saline (PBS), which is commonly used as the liquid in which the cells are kept in dielectrophoretic experiments, and assuming that the voltage used is 10 Vpp, the calculated temperature rise is found to be ca. 34.5°C, considering that the PBS conductivity is 1.6 S/m and the PBS thermal conductivity is 0.58 W/mK (Gielen et al., 2010). Such temperature increase would kill any cells coming in the vicinity of the electrodes. It should be noted that this number is only an order-of-magnitude estimate as the temperature rise depends greatly on the system geometry.

Even assuming that this temperature rise is not problematic for the particles in terms of sample survival, there are other problems associated with it. Since the electric field is nonuniform, the power density and thus the temperature rise is also nonuniform. Variation in the temperature of the liquid causes local gradients in the density, viscosity, permittivity, and conductivity of the medium, and these give rise to forces on the fluid. Two main categories of external forces due to temperature gradients exist: (a) the electrothermal forces, caused by gradients in the permittivity and conductivity, and (b) the buoyancy force caused by density gradients, giving rise to natural convection.

Finally, there is one more category to consider, namely, electroosmotic forces, where the electrical field acts on the free charge in the electrical double layer that forms in the electrode–electrolyte interface.

In the following, a very brief description of these forces will be presented. For more details, the reader is encouraged to consult the appropriate references.

3.3.3.1   Electrothermal Forces

Temperature gradients in the fluid give rise to gradients in the medium conductivity and permittivity. Conductivity gradients produce free volume charges and, therefore, Coulomb forces, while the permittivity gradients give rise to a dielectric force. The general equation describing these forces is given by Ramos et al. (1998)

$$\mathbf{f}_e = \rho_q \mathbf{E} - \frac{1}{2} E^2 \nabla \varepsilon + \frac{1}{2} \nabla \left(\rho_m \frac{\partial \varepsilon}{\partial \rho_m} E^2 \right) \tag{3.34}$$

where $\mathbf{f}_e$ is the electrothermal force per unit volume, ρ_q is the volume charge density, ρ_m the mass density of the medium, and ε the permittivity of the medium. The last term of Equation 3.34 can be ignored for incompressible fluids. This force is frequency dependent, similar to the DEP force, and has a maximum at the electrode surface. Ramos et al. (1998) has provided detailed equations for the force in a special case of parallel infinite length planar electrodes. They also provided an approximate value for the resulting fluid velocity as

$$u_{fluid} \cong |\mathbf{f}_e| \frac{\ell^2}{\eta} \tag{3.35}$$

where ℓ is a characteristic length of the system, and η is the viscosity of the fluid. Using characteristic values for aqueous electrolyte solutions as described in Morgan and Green (2003), the velocity of the fluid for a system with a characteristic length of 10 μm is about 5–50 μm/s at a radial distance of 20 μm away from the electrodes. The calculated velocity is actually of the same order of magnitude as the DEP force on the particles, which means that the electrothermal effects cannot always be ignored. The higher the medium conductivity, the higher will be the liquid velocity due to this effect.

3.3.3.2 Buoyancy Forces

Temperature rise in the fluid due to applied voltage is nonuniform. Therefore, a buoyancy force arises that causes the hotter parts of the fluid to rise and the colder parts of the fluid to fall. This force (in N/m^3) is given by Morgan and Green (2003):

$$\mathbf{f_g} = \frac{\partial \rho_m}{\partial T} \Delta T \cdot \mathbf{g} \tag{3.36}$$

This force is much smaller than electrical forces and is, therefore, often ignored (Ramos et al., 1998). However, contrary to the electrical forces, this force is not frequency dependent and could therefore play a role at certain frequency ranges.

3.3.3.3 Electroosmotic Forces

Electroosmosis is a result of forces acting on the electrical double layer that forms in the electrode–electrolyte interface. Analytical expressions for the fluid movement are difficult to derive for complicated electrode geometries, but competent approximations exist for a system comprised of two parallel and infinite-length electrodes (Gonzalez et al., 2000). In general, the electroosmotic velocity is low at the low- and high-frequency limits but increases to a significant value at a specific frequency dependent on the voltage, electrolyte conductivity, and position across the electrode. The higher the electrolyte conductivity, the higher the frequency at which the electroosmotic velocity has a maximum, but the smaller that maximum velocity is. The maximum velocity is always at the edge of the electrode.

3.3.3.4 Force Comparison

Electroosmosis is a phenomenon that is generally dominant at frequencies around 10 Hz–10 kHz for low-conductivity electrolytes. Although it can be ignored for large particles such as cells, it can completely dominate the movement of submicrometer particles such as small proteins and viruses at low electrolyte conductivities, as the DEP force in these cases is often smaller. For higher electrolyte conductivities, this is not necessarily the case.

Electrothermal forces, on the other hand, are present at all frequencies but result in lower velocities than electroosmosis. Generally, the forces are small for low electrolyte conductivities and low applied potentials, and occur at higher frequencies than electroosmosis. The direction of the fluid flow changes at a frequency around the charge relaxation frequency ($\approx \sigma/\varepsilon$) of the medium. The electrothermal fluid flow can be the dominant effect at high-conductivity media.

3.4 ELECTRODE DESIGN AND DIELECTROPHORESIS SIMULATIONS

When designing a system for the manipulation of biological particles, it is often quite helpful to evaluate the performance of the system before time and effort (and money) is put into its fabrication. Theoretical calculations of the expected forces on the particles as well as the motion of the particle inside the system are often employed in such cases. These can lead to better design of the devices or better utilization of existing devices.

As microfabrication techniques evolve and become generally available, the electrode structures used for generating electric fields for DEP manipulation of particles become more and more complicated, so that analytical expressions for the generated electric fields are difficult or even impossible to obtain. In such cases, numerical simulations are used to generate 2D or 3D graphs of electric field magnitudes, electric field gradients, and even the DEP force on specific particles at specific locations.

Various software solutions exist for doing these simulations, and which one you choose is more a matter of personal preference and level of programming knowledge. One of the most popular and newer software packages for numerical simulations is COMSOL Multiphysics®. First developed as an extra package in MATLAB® for solving differential equations, it has rapidly developed into a stand-alone product with multiple packages that already have all equations preset for practically every physics application you can think of. This means that all you need do is draw your structures, apply the correct boundary conditions (by choosing from a list of predefined options), and press the solve button. No matter which program you choose, some knowledge of the underlying physics is required in order to evaluate the results.

This chapter has so far presented some of the essential theory behind DEP. The theory has concentrated on simple geometrical approximations of biological particles (i.e., homogeneous spheres, prolate ellipsoids, and shells). Quite a few biological particles fall into these categories, so that simple theoretical calculations can be performed. However, more complicated simulations, using advanced features of the various software packages, can be applied for all sorts of particle geometries.

In the following sections, an example of a theoretical calculation involving the movement of homogeneous spherical particles inside a microfluidic channel equipped with electrodes will be described in detail.

3.4.1   CHIP DESIGN AND PROBLEM FORMULATION

In order to study the dielectrophoretic behavior of different cells and possibly separate them, a design has been made that incorporates electrodes inside a microfluidic channel. This design was developed in Demierre et al. (2007) and has been modified here. It is shown in Figure 3.15. One of the features of this design is that the field is almost homogenous along the height of the channel. This means that the force experienced by the particles is practically independent of their position along the channel height.

The purpose of the theoretical calculation is to see how much the electrodes can deflect the path of the cells flowing in the main channel, depending on the applied voltage and the chosen channel dimensions.

In this example, a voltage of 10 V amplitude at a frequency of 5 MHz is applied to the electrodes, and COMSOL Multiphysics is used to calculate the gradient of the square of the electric field in the x, y, and z directions. COMSOL Multiphysics is also used to calculate the flow through the channel (i.e., the velocity of the fluid) when the liquid containing the particles is pumped at 0.576 μL/h. Details of how both the field and its dependence on channel dimensions can be determined analytically are given in Demierre et al. (2007).

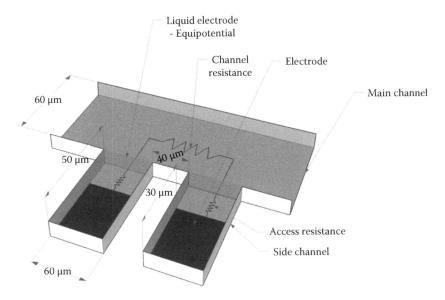

FIGURE 3.15 Schematic of the microfluidic system used in this calculation example. The particles flow in the main channels, and the voltage is applied on the electrodes placed in the side channels. (Figure courtesy of Simon Levinsen.)

The results of the COMSOL simulations are then exported to MATLAB, where a custom-made program calculates the motion of the particle due to DEP forces and fluid drag, as described in Section 3.3. The motion of the particle is then plotted in the geometry for evaluation.

3.4.2 THE GRADIENT

The electric field generated by the electrodes can be calculated by COMSOL. In reality, COMSOL calculates numerically the voltage distribution inside the channel. From that it can then postcalculate the electric field and the gradient of the electric field square, which is proportional to the force. Figure 3.16 shows the logarithm of the gradient of the electric field square 5 and 30 μm above the electrode plane. As mentioned earlier, the gradient does not change much along the height of the channels in the main channel, though it is, of course, larger where it is closer to the electrodes.

3.4.3 PARTICLE MOTION

Having calculated the gradient and the velocity field by COMSOL, it is now easy to theoretically calculate the position of the particle with time based on Equation 3.26. The particle is treated like a point for these calculations, which essentially means that it is assumed that the particle is not in any way distorting the electrical field. This is not really the case for large particles, but the approximation is good enough for particles whose dimensions are smaller than the spatial variation of the gradient.

As Equation 3.26 gives the terminal velocity of the particle, then, for each step the particle takes, one only needs to calculate the velocity and multiply with the time

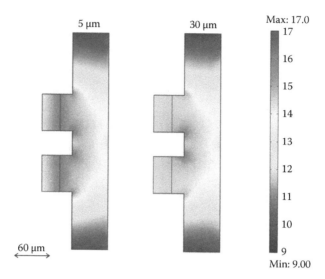

FIGURE 3.16 (See color insert.) The logarithm of the gradient of the square of the magnitude of the electric field inside the channel 5 µm and 30 µm above the electrode plane. The logarithm is plotted instead of the gradient in order to visualize the details better. In this example, the total channel height is 60 µm.

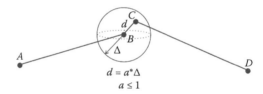

FIGURE 3.17 Calculation of the position of a particle when Brownian motion is taken into account. The displacement AB has been calculated by Equation 3.26. Then, using Equation 3.30, the maximum Brownian displacement is calculated, and the particle is randomly placed at point C distance d from B. The displacement CD is then again calculated by using Equation 3.26.

step to get the particle displacement. If desirable, a pseudo-Brownian motion can be added to this displacement, as shown in Figure 3.17, by using Equation 3.30 to calculate the maximum Brownian displacement for a certain time step. The position of the particle can then be altered from the calculated one to a position within a sphere of radius Δ. Then, the next displacement can be calculated using Equation 3.26, again from the new starting point. Note that if Brownian motion is to be simulated, then the time step chosen for the simulations needs to be small.

In this example, Brownian motion is not taken into account, and the plotted trajectories are only based on Equation 3.26. Trajectories of particles of 1 µm in diameter are plotted. The particles are taken to have a relative permittivity of 100 and a conductivity of 83 mS/m, and are inside a solution of relative permittivity of 43 and conductivity of 0.9 mS/m. The trajectories are plotted for particles starting at 5 µm

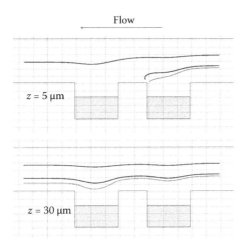

FIGURE 3.18 The calculated trajectories for a 1 μm diameter particle experiencing positive DEP in the system of Figure 3.15. Depending on the position of the particle along the channel height, the trajectories will be different.

and 30 μm above the electrode plane and are shown in Figure 3.18. The particles are experiencing positive DEP and are deflected toward the electrodes.

The observed differences in the trajectories for the two heights are not so much due to the differences in the DEP force, as this design provides rather homogeneous DEP forces along the channel height. But given that the flow in the channel is laminar, the velocity of the flow close to the electrode plane is much smaller than the velocity in the middle of the channel. Therefore, particles closer to the channel walls will be influenced more by the DEP forces and will, therefore, be deflected more toward (or away from, in the case of negative DEP) the electrodes.

3.5 EXAMPLES FROM THE LITERATURE

Most of the literature available on DEP on biological structures concentrates on different types of cells and to a lesser extent on DNA. Primary applications of the method involve cell sorting, for example, live from dead cells (Urdaneta and Smela, 2007; Tai et al., 2007), healthy from cancer cells (Becker et al., 1994; Yang et al., 1999a; An et al., 2008), sorting between different types of cells (Gasperis et al., 1999;Yang et al., 2000a; Wang et al., 2000; Burgarella et al., 2010; Yang et al., 2010; Flanagan et al., 2008), size sorting of DNA (Regtmeier et al., 2007), dielectric properties mapping (Regtmeier et al., 2007; Menachery and Pethig, 2005; Fatoyinbo et al., 2008; Tuukkanen et al., 2007), etc., and biological structure trapping (Suehiro and Pethig, 1998; Hölzel et al., 2005; Tuukkanen et al., 2007; Koklu et al., 2010; Yeo et al., 2009; Gallo-Villanueva et al., 2009; Castillo et al., 2008), for example, in order to create cell clusters (Pethig et al., 2008). Subcellular components (Moschallski et al., 2010), viruses (Ermolina et al., 2006), and bacteria (Moncada-Hernandez and Lapizco-Encinas, 2010; Kim and Soh, 2009) can also be manipulated by dielectrophoresis.

A summary of some of the literature used in this paper is found in Table 3.2 in terms of authors, method used, and biological structure used. To name but a few examples, Wang et al. (2000) used interdigitated electrodes at the bottom of a chamber to enrich leukocytes from blood and separate T- and B-lymphocytes, monocytes, and granulocytes. Differentiation of human leukocyte subpopulations is very important, for example, for the analysis of the functionality of the different leukocytes in immune responses (Yang et al., 2000a). As most biological structures survive in rather high conductivity solutions, Marcus et al. (Marcus et al., 2007) have studied how DEP works in such environments using carbon nanofibers. When properly functionalized, these nanowires can work as biologically actuated electrical switches. The authors demonstrate that DEP can work in saline solutions when care is taken to keep the frequency above a certain critical limit. In a recent article, Pethig et al. (2008) create insulinoma cell clusters of about 1000 cells, which they term *pseudo-islets*, as the structures resemble in size the islets of Langerhans found in the pancreas, where they regulate insulin levels in the blood. The authors hope to be able to uncover how cells work together in complexes in the body.

More recently, Koklu et al. (2010) used a combination of negative DEP and electrothermal flow (ETF) in order to trap bacterial spores on specified locations from high-conductivity media such as orange juice. Their system allows further examination of the spores with assays in the location they are trapped. In Valero et al., (2010) the authors use multiple-frequency DEP in a configuration that allows different cells to reach individual equilibrium positions inside a microfluidic channel and thus be spatially separated at the outlet. They manage to separate dead from live yeast cells, dividing undividing yeast cells as well as healthy red blood cells (RBCs) from infected RBCs.

In Burgarella et al. (2010), the authors have developed a system that combines several modules, each for a different purpose. Thus, they are able to separate cells (red and white blood cells), trap them for optical observations, and sort the different subpopulations of white blood cells. In Moschallski et al. (2010), a system has been developed with electrodes at the bottom and top of a microfluidic channel that is capable of providing mitochondria fractions six times less contaminated by other intracellular components than state-of–the-art technologies. In Moncada-Hernandez and Lapizco-Encinas (2010), the authors use a method called *insulator-based DEP* (iDEP) to separate *E. coli* bacteria and yeast cells. Insulator-based DEP is similar to DEP with the exception that, instead of electrodes inside microfluidic channels, there are physical obstacles made of insulating material that constrict not only the flow but also the electric field lines, thus creating inhomogeneous fields. The advantage is that the biological objects never come into contact with the high voltage used to generate the electrical fields, and the fabrication is often also cheaper as no electrodes are needed. The same technique is applied in Gallo-Villanueva et al. (2009) to trap DNA particles. In Kim and Soh (2009), a combination of dielectrophoretic and magnetic sorting is used in the same chip to separate three distinct clones of *E. coli*.

From the foregoing text, it is evident that DEP has a huge potential in manipulating biological structures and has the advantage of being compatible with standard microelectronics foundry technologies, allowing the integration of the DEP

TABLE 3.2
Summary of a Fraction of the Literature Available on the Dielectrophoretic Manipulation of Biological Objects

Manipulation Technique	Manipulated Nanoobject	Reference
DEP	DNA, proteins, viruses, bacteria, chromosomes, etc.	(Lapizco-Encinas and Rito-Palomares, 2007) (Review article)
DEP	DNA, proteins, viruses, bacteria, chromosomes, etc.	(Pethig, 2010) (Review article)
DEP	DNA	(Regtmeier et al., 2007)
DEP	Protein molecules (R-phycoerythrin)	(Hölzel et al., 2005)
DEP	DNA, thiol-modified DNA	(Tuukkanen et al., 2007)
DEP	Yeast pellets and red blood cells	(Fatoyinbo et al., 2008)
DEP	Cardiac myocytes	(Yang and Zhang, 2007)
DEP	Polystyrene particles	(Chen et al., 2007)
DEP	Live and dead yeast cells	(Li et al., 2007)
DEP	Polystyrene particles, red blood cells, *E. coli*	(Jung and Kwak, 2007)
DEP	Cow Pea Mosaic Virus (CPMV) and Tobacco Mosaic Virus (TMV)	(Ermolina et al., 2006)
DEP tweezer	Yeast cell	(Lee et al., 2007)
DEP	T-cells, monocytes	(Pethig et al., 2004)
DEP	Jurkat T-cells, HL60 leukemia cells	(Menachery and Pethig, 2005)
DEP	Human breast cancer cells, T-lymphocytes CD34+ hematopoietic stem cells, leukocytes	(Wang et al., 2000)
DEP	Human leukocytes	(Yang et al., 2000a)
2D DEP	MDA-435, HL-60, and DS19, PBMN cells	(Gasperis et al., 1999)
DEP	HL-60 cells and blood	(Becker et al., 1994)
DEP/G-FFF	MDA-435, blood cells	(Yang et al., 1999a)
DEP	Protoplasts	(Suehiro and Pethig, 1998)
DEP	BETA-TC-6, INS-1 cells	(Pethig et al., 2008)
DEP	Nanowires	(Marcus et al., 2007)
DEP	MCF10A, MCF7 cells	(An et al., 2008)
DEP	Human lung cancer cells	(Tai et al., 2007)
DEP (multiple frequency)	Yeast cells	(Urdaneta and Smela, 2007)
DEP, electrowetting	Neuro-2a cells, polystyrene beads	(Fan et al., 2008)
DEP	Silica microspheres	(Zou et al., 2008)
DEP and ETF	Bacterial spores (*Bacillus subtilis* and *Clostridium sporogenes*)	(Koklu et al., 2010)

(Continued)

TABLE 3.2   (CONTINUED)

Summary of a Fraction of the Literature Available on the Dielectrophoretic Manipulation of Biological Objects

Manipulation Technique	Manipulated Nanoobject	Reference
DEP	RBC, WBC	(Burgarella et al., 2010)
DEP	Mitochondria	(Moschallski et al., 2010)
DEP	HTC116 colon cancer cells, Human Embryonic	
Kidney 293 cells (HEK 293), and *Escherichia coli* (*E. coli*) bacteria	(Yang et al., 2010)	
DC iDEP	*E. coli* and yeast cells	(Moncada-Hernandez and Lapizco-Encinas, 2010)
DEP and magnetic separation	*E. coli* MC1061	(Kim and Soh, 2009)
DEP and capillary action	Extracellular DNA	(Yeo et al., 2009)
DEP (multiple frequency)	Yeast cells, RBCs	(Valero et al., 2010)
DEP	Neuron/stem precursor cells, differentiated neurons, and differentiated astrocytes	(Flanagan et al., 2008)
DC iDEP	Linear DNA particles (pET28b)	(Gallo-Villanueva et al., 2009)
DEP	Peptide nanotubes	(Castillo et al., 2008)

devices with transduction, readout, signal processing, and communications circuitry. However, DEP is also notorious for the rather limited precision control it offers, as well as the difficulties involved with the *in situ* manufacturing process monitoring. These two drawbacks have limited the yield and future commercialization of DEP systems (Sitti, 2007), although, in recent years, a large number of patents have been filed on applications ranging from device fabrication to biomedical diagnostics (Pethig, 2010).

FURTHER READING

This chapter has presented the essential theoretical knowledge on DEP and a few of the potential applications that can be found in the literature. The review is by no means complete, and the reader is therefore encouraged to read the recent excellent review by Pethig (2010) that contains more than 250 references to recent and older work. Moreover, for in-depth theoretical considerations on AC electrokinetic phenomena in general, the reader is encouraged to consult the textbook by Morgan and Green (Morgan and Green, 2003) and the references therein.

REFERENCES

An, J., Lee, J., Kim, Y., Kim, B., and Lee, S. (2008). Analysis of cell separation efficiency in dielectrophoresis-activated cell sorter. *3rd IEEE International Conference on Nano/ Micro Engineered and Molecular Systems.* Sanya, China.

Becker, F. F., Wang, X.-B., Huang, Y., Pethig, R., Vykoukal, J., and Gascoyne, P. R. C. (1994). The removal of human leukaemia cells from blood using interdigitated microelectrodes. *Journal pf Physics D: Applied Physics,* 27, 2659–2662.

Burgarella, S., Merlo, S., Dell'anna, B., Zarola, G., and Bianchessi, M. (2010). A modular micro-fluidic platform for cells handling by dielectrophoresis. *Microelectronic Engineering,* 87, 2124–2133.

Castillo, J., Tanzi, S., Dimaki, M., and Svendsen, W. (2008). Manipulation of self-assembly amyloid peptide nanotubes by dielectrophoresis. *Electrophoresis,* 29, 5026–5032.

Chen, D. F., Du, H., and Li, W. H. (2007). Bioparticle separation and manipulation using dielectrophoresis. *Sensors and Actuators A,* 133, 329–334.

Demierre, N., Braschler, T., Linderholm, P., Seger, U., Van Lintel, H., and Renaud, P. (2007). Characterization and optimization of liquid electrodes for lateral dielectrophoresis. *Lab on a Chip,* 7, 355–365.

Ermolina, I., Milner, J., and Morgan, H. (2006). Dielectrophoretic investigation of plant virus particles: Cow Pea Mosaic Virus and Tobacco Mosaic Virus. *Electrophoresis,* 27, 3939–3948.

Fan, S.-K., Huang, P.-W., Wang, T.-T., and Peng, Y.-H. (2008). Cross-scale electric manipulations of cells and droplets by frequency-modulated dielectrophoresis and electrowetting. *Lab on a Chip,* 8, 1325–1331.

Fatoyinbo, H. O., Hoettges, K. F., and Hughes, M. P. (2008). Rapid-on-chip determination of dielectric properties of biological cells using imaging techniques in a dielectrophoresis dot microsystem. *Electrophoresis,* 29, 3–10.

Flanagan, L. A., Lu, J., Wang, L., Marchenko, S. A., Jeon, N. L., Lee, A. P., and Monuki, E. S. (2008). Unique dielectric properties distinguish stem cells and their differentiated progeny. *Stem Cells,* 26, 656–665.

Gallo-Villanueva, R. C., Rodriguez-Lopez, C. E., Diaz-De-La-Garza, R. I., Reyes-Betanzo, C., and Lapizco-Encinas, B. H. (2009). DNA manipulation by means of insulator-based dielectrophoresis employing direct current electric fields. *Electrophoresis,* 30, 4195–4205.

Gasperis, G. D., Yang, J., Becker, F. F., Gascoyne, P. R. C., and Wang, X.-B. (1999). Microfluidic cell separation by 2-dimensional dielectrophoresis. *Biomedical Microdevices,* 2, 41–49.

Gielen, F., Pereira, F., Demello, A. J., and Edel, J. B. (2010). High-resolution local imaging of temperature in dielectrophoretic platforms. *Analytical Chemistry,* 82, 7509–7514.

Gimsa, J. (2001). A comprehensive approach to electro-orientation, electrodeformation, dielectrophoresis, and electrorotation of ellipsoidal particles and biological cells. *Bioelectrochemistry,* 54, 23–31.

Gonzalez, A., Ramos, A., Green, N. G., Castellanos, A., and Morgan, H. (2000). Fluid flow induced by nonuniform ac electric fields in electrolytes on microelectrodes. II. A linear double-layer analysis. *Physical Review E,* 61, 4019–4028.

Hölzel, R., Calander, N., Chiragwandi, Z., Willander, M., and Bier, F. F. (2005). Trapping single molecules by dielectrophoresis. *Physical Review Letters,* 95, 128102.

Jung, J.-Y. and Kwak, H.-Y. (2007). Separation of microparticles and biological cells inside an evaporating droplet using dielectrophoresis. *Analytical Chemistry,* 79, 5087–5092.

Kim, U. and Soh, H. T. (2009). Simultaneous sorting of multiple bacterial targets using integrated Dielectrophoretic-Magnetic Activated Cell Sorter. *Lab on a Chip,* 9, 2313–2318.

Koklu, M., Park, S., Pillai, S. D., and Beskok, A. (2010). Negative dielectrophoretic capture of bacterial spores in food matrices. *Biomicrofluidics*, 4, 15.

Lapizco-Encinas, B. H. and Rito-Palomares, M. (2007). Dielectrophoresis for the manipulation of nanobioparticles. *Electrophoresis*, 28, 4521–4538.

Lee, K., Kwon, S. G., Kim, S. H., and Kwak, Y. K. (2007). Dielectrophoretic tweezers using sharp probe electrode. *Sensors and Actuators A*, 136, 154–160.

Li, Y., Dalton, C., Crabtree, J., Nilsson, G., and Kaler, K. V. I. S. (2007). Continuous dielectrophoretic cell separation microfluidic device. *Lab on a Chip*, 7, 239–248.

Marcus, M. S., Shang, L., Li, B., Streifer, J. A., Beck, J. D., Perkins, E., Eriksson, M. A., and Hamers, R. J. (2007). Dielectrophoretic manipulation and real-time electrical detection of single-nanowire bridges in aqueous saline solutions. *Small*, 3, 1610–1617.

Menachery, A. and Pethig, R. (2005). Controlling cell destruction using dielectrophoretic forces. *IEE Proceedings Nanobiotechnology*, 152, 145–149.

Moncada-Hernandez, H. and Lapizco-Encinas, B. H. (2010). Simultaneous concentration and separation of microorganisms: insulator-based dielectrophoretic approach. *Analytical and Bioanalytical Chemistry*, 396, 1805–1816.

Morgan, H. and Green, N. G. (2003). *AC Electrokinetics: Colloids and Nanoparticles*. Research Studies Press Ltd. Baldock, United Kingdom.

Moschallski, M., Hausmann, M., Posch, A., Paulus, A., Kunz, N., Duong, T. T., Angres, B., Fuchsberger, K., Steuer, H., Stoll, D., Werner, S., Hagmeyer, B., and Stelzle, M. (2010). MicroPrep: Chip-based dielectrophoretic purification of mitochondria. *Electrophoresis*, 31, 2655–2663.

Pethig, R. (2010). Review article-dielectrophoresis: Status of the theory, technology, and applications. *Biomicrofluidics*, 4, 35.

Pethig, R., Lee, R. S., and Talary, M. S. (2004). Cell physiometry tools based on dielectrophoresis. *JALA*, 9, 324–330.

Pethig, R., Menachery, A., Heart, E., Sanger, R. H., and Smith, P. J. S. (2008). Dielectrophoretic assembly of insulinoma cells and fluorescent nanosensors into three-dimensional pseudo-islet constructs. *IET Nanobiotechnology*, 2, 31–38.

Ramos, A., Morgan, H., Green, N. G., and Castellanos, A. (1998). Ac electrokinetics: a review of forces in microelectrode structures. *Journal of Physics D—Applied Physics*, 31, 2338–2353.

Regtmeier, J., Duong, T. T., Eichhorn, R., Anselmetti, D., and Ros, A. (2007). Dielectrophoretic Manipulation of Dna: Separation and Polarizability. *Analytical Chemistry*, 79, 3925–3932.

Sitti, M. (2007). Microscale and nanoscale robotics systems [Grand Challenges of Robotics]. *IEEE Robotics and Automation Magazine*, 14, 53–60.

Suehiro, J. and Pethig, R. (1998). The dielectrophoretic movement and positioning of a biological cell using a three-dimensional grid electrode system. *Journal of Physics D—Applied Physics*, 31, 3298–3305.

Tai, C.-H., Hsiung, S.-K., Chen, C.-Y., Tsai, M.-L., and Lee, G.-B. (2007). Automatic microfluidic platform for cell separation and nucleus collection. *Biomedical Microdevices*, 9, 533–545.

Tuukkanen, S., Kuzyk, A., Toppari, J. J., Häkkinen, H., Hytönen, V. P., Niskanen, E., Rinkiö, M., and Törmä, P. (2007). Trapping of 27 bp-8 kbp DNA and immobilization of thiol-modified DNA using dielectrophoresis. *Nanotechnology*, 18, 295204.

Urdaneta, M. and Smela, E. (2007). Multiple frequency dielectrophoresis. *Electrophoresis*, 28, 3145–3155.

Valero, A., Braschler, T., Demierre, N., and Renaud, P. (2010). A miniaturized continuous dielectrophoretic cell sorter and its applications. *Biomicrofluidics*, 4, 9.

Wang, X.-B., Yang, J., Huang, Y., Vykoukal, J., Becker, F. F., and Gascoyne, P. R. C. (2000). Cell Separation by Dielectrophoretic Field-flow-fractionation. *Analytical Chemistry*, 72, 832–839.

Yang, F., Yang, X. M., Jiang, H., Bulkhaults, P., Wood, P., Hrushesky, W., and Wang, G. R. (2010). Dielectrophoretic separation of colorectal cancer cells. *Biomicrofluidics*, 4, 13.

Yang, J., Huang, Y., Wang, X.-B., Becker, F. F., and Gascoyne, P. R. C. (1999a). Cell separation on microfabricated electrodes using dielectrophoretic/gravitational field-flow fractionation. *Analytical Chemistry*, 71, 911–918.

Yang, J., Huang, Y., Wang, X. J., Wang, X. B., Becker, F. F., and Gascoyne, P. R. C. (1999b). Dielectric properties of human leukocyte subpopulations determined by electrorotation as a cell separation criterion. *Biophysical Journal*, 76, 3307–3314.

Yang, J., Huang, Y., Wang, X.-B., Becker, F. F., and Gascoyne, P. R. C. (2000a). Differential analysis of human leucocytes by dielectrophoretic field-flow-fractionation. *Biophysical Journal*, 78, 2680–9.

Yang, J., Huang, Y., Wang, X. B., Becker, F. F., and Gascoyne, P. R. C. (2000b). Differential analysis of human leukocytes by dielectrophoretic field-flow-fractionation. *Biophysical Journal*, 78, 2680–2689.

Yang, M. and Zhang, X. (2007). Electrical assisted patterning of cardiac myocytes with controlled macroscopic anisotropy using a microfluidic dielectrophoresis chip. *Sensors and Actuators A*, 135, 73–79.

Yeo, W. H., Chung, J. H., Liu, Y. L., and Lee, K. H. (2009). Size-specific concentration of DNA to a nanostructured tip using dielectrophoresis and capillary action. *Journal of Physical Chemistry B*, 113, 10849–10858.

Zou, Z., Lee, S., and Ahn, C. H. (2008). A polymer microfluidic chip with interdigitated electrodes arrays for simultaneous dielectrophoretic manipulation and impedimetric detection of microparticles. *IEEE Sensors Journal*, 8, 527–535.

4 Optical Manipulation Techniques

Kirstine Berg-Sørensen
Department of Physics, Technical
University of Denmark, Lyngby

CONTENTS

4.1   INTRODUCTION

The idea to use light to manipulate microscopic or nanoscopic objects in a laboratory setting was formulated very soon after the advent of the laser. The field was pioneered by Arthur Ashkin, who along with his coworkers managed to first trap dielectric spheres and later living cells (Ashkin and Dziedzic, 1987; Ashkin et al., 1986; Ashkin et al., 1987). Historically, optical manipulation of and within living objects progressed simultaneously—and in some cases, in the same laboratories—with the early investigations to use light to trap and cool atomic species. Excellent accounts of the early history of the use of light to trap and manipulate biological objects may be found in the Nobel Prize acceptance speech of Steven Chu (Chu, 1998) or in an inaugural lecture for the National Academy of Sciences of the United States by Arthur Ashkin (Ashkin, 1997).

Since the mid-1990s, the technique of optical trapping of biological samples, both *in vitro* and, lately, also *in vivo*, has been developed to become a method for high-precision force spectroscopy. Among its virtues is that it is almost noninvasive and nondestructive; it is naturally applied in aqueous surroundings and can be sterile.

The most standard optical manipulation technique, single-beam optical tweezers, can be constructed within a research-grade optical microscope, already available in many biological research laboratories. Furthermore, it may be combined with fluorescence detection and also designed within a confocal microscope. For users not inclined to set up their own equipment, commercial systems are also available.

For investigations of biological samples, optical traps may be constructed simply to manipulate single molecules, organelles within living cells, or (small) single cells, but the main use of optical tweezers in the research literature is as a force-scope, applying or measuring forces in the pN range, with standard position resolution in the nanometer range—forces and distances that are relevant for biological systems. The trapping laser is typically a CW laser chosen in the infrared, in order to minimize absorption and thereby heating of the biological samples. In one study of living cells, the wavelengths 830 nm and 970 nm have been shown to be particularly suitable, that is, with minimum absorption (Neuman et al., 1999). Depending on the exact application required, the intensities of small and inexpensive diode lasers may even be sufficient, although many dedicated optical tweezers laboratories use solid-state lasers such as Nd:YAG or Nd:YVO4 lasers with a wavelength of 1064 nm. Apart from absorption in the biological sample, heating due to absorption in the water may be an issue to consider (Peterman et al., 2003a). However, as the intensities needed to exert sufficient forces are typically in the hundreds of milliwatts regime, heating and absorption very rarely become a problem.

4.2   OPTICAL FORCES

The phenomenon of optical trapping can be understood from an analysis of the forces exerted by the laser light on the object one wants to trap. The fundamental physics is similar to the physics of dielectrophoresis (discussed in Chapter 3), but with electric fields varying at the high frequencies corresponding to light waves. For manipulation by light waves to be possible, first of all, the refractive index of the particle to be trapped or manipulated must be sufficiently different from that of the surroundings. The size and direction of the force, and how the force should be calculated, now depends on the size of the object relative to the wavelength of the light. In most theoretical accounts of the optical forces, the object to be manipulated is assumed spherical, or simply, in cases where the size of the object is negligible in comparison with the wavelength of the light, approximated by a point dipole. In this latter case, the calculation of the force is based on electromagnetic theory—on the contrary, when the object is large as compared to the wavelength of the light, the forces may be calculated from a ray-optics description.

In single-molecule biophysics, typically a spherical probe particle that has the right properties to be trapped by light is attached by biochemical means to the biological molecule of interest. Most often, micron-sized dielectric spheres have been used for this purpose, but nanometer-sized metallic particles have also been suggested, as reviewed in more detail by Dienerowitz et al. (2008). The use of spherical handles is particularly useful when force-scope applications are needed as the force on a spherical object is better characterized than the forces on a molecule (e.g., DNA).

In *in vivo* studies using optical tweezers, naturally occurring objects with appropriate optical properties such as lipid granuli (Sacconi et al., 2005; Tolic-Nørrelykke et al., 2004b) have been trapped and manipulated and, for example, applied to move the cell nucleus in *S. pombe* (Tolic-Nørrelykke et al., 2005). In other studies, nanometer-sized metal particles have been injected in living cells for subsequent manipulation by optical tweezers (Hansen and Oddershede, 2005).

The early literature on optical trapping in connection with biological samples also describes the trapping by single-beam traps of single cells such as bacteria (Ashkin and Dziedzic, 1987; Ashkin et al., 1987). More recently, mechanical properties of, for example, red blood cells have been investigated by means of optical trapping both with single- or dual-beam optical tweezers (Huang et al., 2008) or with a trap constructed by two counterpropagating light beams (Guck et al., 2001; Guck et al., 2000). In Sections 4.8 and 4.9, these applications will be described in more detail. For now, we note that in studies where force spectroscopy is sought, the object actually trapped and manipulated by the optical trap, when listed in order of ascending size, ranges from naturally occurring objects in a cell over dielectric microspheres to metallic nanoparticles and single quantum dots (Jauffred et al., 2008).

When the size of the trapped object, r, is very small compared to the wavelength of the light, λ, we will assume that the trapped object can be approximated by a (point) dipole, with a dipole moment induced by the electric field from the laser light, and understand the existence of trapping forces via the dipole approximation for the interaction between light and matter. This approach is taken for metallic nanoparticles, quantum dots, and, for that matter, atoms. For the smallest particles in this size range, quantitative force calculations are based on Rayleigh scattering coefficients, whereas Lorenz–Mie scattering is applied to determine forces on slightly larger particles (Dienerowitz et al. 2008). Detailed calculations that are in good agreement with experimental measurements for dielectric spheres of size r of order just below λ have been presented by Rohrbach (2005).

The force acting on an object positioned in the electric field of the laser light is composed of a scattering force in the direction of propagation of the laser light and a gradient force that is also known as the *dipole force*. Let us for a moment consider the latter force. In quantitative terms, the induced-dipole moment of the object to be trapped can be expressed as

$$\mathbf{p} = \alpha \mathbf{E} \tag{4.1}$$

where α is the polarizability of the object to trap. The energy of interaction of this induced dipole in the electric field is

$$V_{dip}^{ind} = -\int_0^E \mathbf{p} \cdot d\mathbf{E} = -\frac{1}{2} \alpha \, \mathbf{E}^2 \tag{4.2}$$

from which the force may be found

$$\mathbf{F}_{dip} = -\nabla V_{dip}^{ind} = \frac{\alpha}{2} \nabla \mathbf{E}^2 \tag{4.3}$$

The polarizability, α, is related to the relative index of refraction $m \equiv n_{bead}/n_{medium}$, where n_{medium} and n_{bead} are the refractive indices of medium and bead to trap. For a sphere of radius α,

$$\alpha = n_{medium}^2 a^3 \left(\frac{m^2 - 1}{m^2 + 2} \right)$$

(4.4)

We immediately note that the polarizability is positive when $m>1$, or $n_{bead}>n_{medium}$, and as a result, the force points in the direction of increasing intensity of the light.

If, on the contrary, $r \gg \lambda$, we will describe and calculate forces as based on a ray optics picture.

Further, we shall in this section assume the trapped object to be spherical. The ray optics approach is taken for (large) dielectric spheres or cells. If we further simplify and consider refraction only, the situation for a dielectric sphere in a focused laser beam with a Gaussian intensity profile is illustrated in Figure 4.1. In this sketch, we consider the effect of two rays only, one from the intense central part of the laser beam and one from the less intense tail region of the intensity profile. The refractive index of the bead, n_{bead}, is higher than that of the surrounding medium, n_{medium}. We see that, because of conservation of momentum, the net effect of these two rays impinging on the dielectric sphere positioned to the left of the intensity maximum

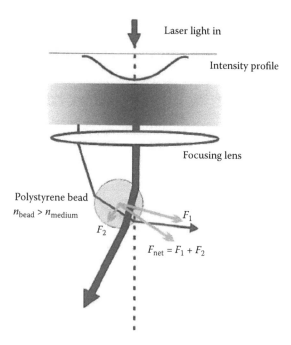

FIGURE 4.1 Sketch to illustrate the forces arising when rays from the focused Gaussian beam profile is refracted on a dielectric sphere. The resulting force from the two rays drawn has a component in the direction of propagation (scattering force) and a component in the direction of the gradient of intensity (gradient force).

is to pull the bead to the right, toward the center of the beam. In the situation in the figure, there is also a component of the force in the direction of propagation of light, acting to push the bead downward. However, if the focusing effect of the lens is sufficient, there will also be a component of the gradient force in the direction of propagation of the light. When we sum up the contributions to the total force, resulting from conservation of momentum when rays are refracted, the forces acting on a dielectric bead positioned below the focus would be seen to pull the bead upward, against the direction of propagation. Thus, including rays from all parts of the intensity profile, and with a lens of sufficiently high NA, we can argue that the focused laser beam acts as a three-dimensional trap for dielectric spheres of refractive indices higher than the surrounding medium.

As will be discussed in more detail in Section 4.7, the optical trapping potential is, to a very good approximation, harmonic. Yet, for very large excursions away from the minimum of the trapping potential, it is important to investigate and account for nonharmonicity (Richardson et al., 2008). On the other hand, as long as the excursions are small enough that the trapping potential stays harmonic, the trapping force can be described as $F_{OT} = -\kappa x$. Thus, force spectroscopy experiments require an independent and precise determination of the spring constant κ and a precise measurement of the position x. Different means of position detection are discussed in Section 4.6. As most experimental applications have $r \sim \lambda$, where calculations of κ are cumbersome, typically κ is determined experimentally, by calibration, as discussed in much more detail in Section 4.7.

4.3 SINGLE-BEAM OPTICAL TWEEZERS

Optical trapping may, in practice, involve one or two laser beams. The most widely used optical trap is known as *optical tweezers* and is constructed from a tightly focused single laser beam. It is convenient to let the laser light into a research-grade optical microscope by means of a dichroic mirror, to make the light path of the trapping laser coincide with the normal white light of the microscope used for viewing the sample as illustrated in Figure 4.2. The trapping laser is then focused by the microscope objective, chosen to have a high numerical aperture, NA > 1. Thus, the focal plane of the microscope to a very good approximation coincides with the trapping plane of the optical tweezers. A detailed protocol for the practical implementation of the experimental setup is available in *Nature Protocols* (Lee et al., 2007).

When comparing the actual forces in such an experimental setup to theoretical investigations, it is important to account for spherical aberrations. These appear because of a mismatch in refractive indices when light travels from the objective through an immersion fluid, then through the bottom of the sample chamber typically made of a microscope slide, and finally into the aqueous medium in which the sample is trapped. With oil immersion objectives, depending on the numerical aperture of the focusing lens, the optimal distance between the inside bottom surface of the sample chamber and the trap is typically on the order of order 10 μm. The exact optimal trapping depth, and strength, may be tuned by a careful choice of refractive index of the immersion liquid (Reihani and Oddershede, 2007). With a water immersion objective, typically the numerical aperture is lower but, at the same time, the

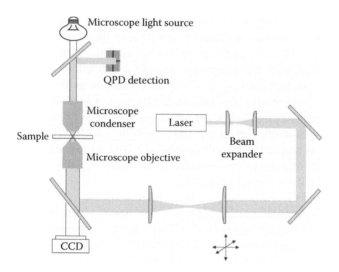

FIGURE 4.2 Sketch of an optical tweezers setup, here with brightfield Koehler illumination. The optics in the left part of the figure, that is, along the light path from the white-light source, may conveniently be chosen as the parts of a research-grade optical microscope. In the right part of the figure, the laser beam is first expanded and then aligned by two mirrors into a set of telescope lenses that allow for steering of the beam in the sample plane.

distance to the surface can be somewhat larger, on the order of order 100 μm, still with a strong and efficient harmonic optical trap.

4.4 TWO-BEAM OPTICAL TRAPS

The very first optical traps suggested and demonstrated (Ashkin, 1970; Roosen, 1977) were based on two counterpropagating laser beams, in which trapping occurs at the position where scattering forces from the two beams balance each other. With slightly diverging laser beams and variation of light intensity, a setup may be constructed that allows for stretching of the trapped object. This setup is known as the *optical stretcher* (Guck et al., 2000; Guck et al., 2001). In such a setup, typically, single-mode optical fibers are applied for delivery of the light. If such a setup is combined with an optical microscope, it is worth noting that one important difference as compared to the optical tweezers setup is that the direction of propagation of the light typically and conveniently lies within the viewing plane. This allows, for example, for direct viewing of formation of "optical matter" (Burns et al., 1989) when several particles are trapped along the direction of propagation of the light. This phenomenon is illustrated in Figure 4.3. The "crystallization" is a result of interaction between the individual induced dipoles, in the images here illustrated with red blood cells. The optical stretcher is conveniently combined with microfluidics, and has in this form been coined the μOS (Lincoln et al., 2007). A very useful design of this type of trap, in which the fiber ends are not in direct contact with the sample, is sketched in Figure 4.5.

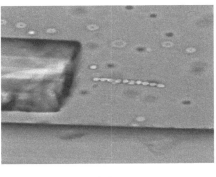
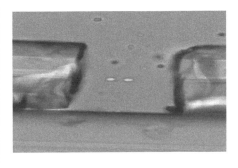

(A)   (B)

FIGURE 4.3 (A) Several red blood cells aligned by optical forces from one fiber balanced by fluid flow. The fiber is a single-mode fiber, and the laser has a wavelength of 980 nm. The cross-sectional diameter of the fiber is 125 μm. (B) Two red blood cells trapped between two optical fibers. Fiber dimensions and wavelength as in the left part of the figure.

The fact that two opposing, slightly diverging, laser beams may stretch rather than compress a trapped object follows from a consideration of momentum conservation. As for the single-beam optical tweezers, the object to trap has a refractive index n_{object} that is larger than the refractive index n_{medium} of the surrounding liquid medium. In the liquid medium, the momentum of a ray of light with energy E is $p_{medium} = n_{medium}E/c$, where c is the speed of light in vacuum. Consider a cube of refractive index $n_{object} > n_{medium}$ and one laser beam incident perpendicular to the surface of the cube, with a reflection coefficient equal to R; $R \ll 1$. As $n_{object} > n_{medium}$, the light that is transmitted gains momentum, and due to momentum conservation, a force F_{front} opposing the propagating light therefore acts on the front surface of the cube,

$$F_{front} = \left(n_{medium} - (1-R)n_{object} + Rn_{medium}\right)\frac{P}{c} \qquad (4.5)$$

Here, P is the total light power. Similarly, at the backside of the cube, the light that is transmitted loses momentum, and momentum conservation results in a force pointing in the direction of propagation of the light, F_{back},

$$F_{back} = \left(n_{object} - (1-R)n_{medium} + Rn_{object}\right)(1-R)\frac{P}{c} \qquad (4.6)$$

The net translatory force on the cube—corresponding to the scattering force—is $F_{total} = F_{back} - F_{front}$, but the surface forces also act to stretch the cube with a force equal to $F_{stretch} = (F_{front}+F_{back})/2$. With a power of $P = 500$ mW, refractive indices of $n_{medium} = 1.33$ and $n_{object} = 1.45$, and reflection coefficient of $R = 0.002$, the scattering force is of order 20 pN, whereas the surface force is 10 times larger, of order 200 pN (Gluck et al, 2000). The forces are illustrated in Figure 4.4.

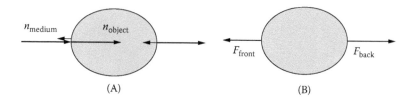

(A)   (B)

FIGURE 4.4 Transmission and reflection of rays of light (A) and the resulting forces (B) acting on the vertical surfaces of an ellipsoid of refractive index n_{object}, positioned in a liquid medium of refractive index n_{medium}. In (A), thick, long arrows illustrate the incoming and transmitted rays, whereas short arrows illustrate the reflected rays.

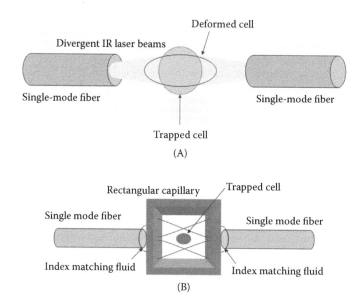

FIGURE 4.5 The optical stretcher. In (A), the principle of the optical stretcher is sketched. At low laser power, the cell is trapped and subsequently stretched as a result of an increase in laser power. In (B), a practical implementation of the μOS, where the optical fibers are not in direct contact with the sample, is sketched. The microfluidic chamber is simply a rectangular capillary in borosilicate glass. Fluid flow is in the direction (say) out of the plane of the paper. This design avoids that the fiber tips get contaminated with contents of the sample in contrast to the trapping illustrated in Figure 4.3.

In the literature, the name *dual-beam* optical traps is more often used to describe experimental setups with two optical tweezers constructed from two copropagating lasers, with focal points in the same plane perpendicular to the axis of propagation, but slightly displaced from one another. An example of such a setup is sketched in Figure 4.10A, illustrating experiments to measure the action of RNA polymerase to below Angstrom resolution (Abbondanzieri et al., 2005). This high resolution may be achieved in a configuration where one trap is very strong, whereas the other has a stiffness constant of the same order as the polymer. With its better precision, the double-optical tweezers assay has thereby been demonstrated to be advantageous over both a setup with a micropipette and one single-beam optical trap (cf. Figure 4.10C), and a setup with one single-beam optical trap and tethered DNA. Similarly, for investigations of nonprocessive molecular motors (see Section 4.8 and Figure 4.10B), the dual-beam copropagating optical tweezers is the assay to choose.

These assays are discussed more thoroughly in Section 4.8.

4.5 OTHER DESIGNS OF OPTICAL TRAPS

Multiple optical traps may be constructed by beam-sharing using acousto-optic or electro-optic deflector systems. If more than a couple of optical traps are wanted, a powerful procedure is by the application of holographic methods (Grier, 2003), and multiple dynamic trap systems can be obtained through the use of spatial light modulators (SLMs) (Eriksen et al., 2002; Reicherter et al., 1999). With SLMs, the beam may be configured to be focused at a specific depth in turbid media, giving hopes for the construction of optical traps of use even in biological systems at the physiological scale.

In addition, optical traps may be constructed from light beams with an angular momentum in Laguerre–Gaussian or Bessel eigenmodes. These types of light beams allow for rotational dynamics to be studied and/or induced, and they have been suggested as a means of constructing light-based tools for microfluidic purposes (pumping, sorting).

Optical traps constructed holographically, with SLMs or with beams of angular momentum, have, with biological samples, mainly been applied to cellular systems, in contrast to the many single-molecule experiments using standard optical tweezers. For further details, many of these applications are described in a recent book chapter (Dholakia and Lee, 2008) and will not be covered further here.

4.6 POSITION DETECTION

Position detection may either take place directly through imaging or through various nonimaging methods.

Detection that involves imaging methods naturally contains very detailed information on the experimental situation and the observed dynamics. Nevertheless, imaging methods using normal CCD cameras suffer from low temporal resolution. High-speed cameras are a natural alternative choice and, with the continuous development of larger CMOS chips and better electronics along with reduced costs of high-speed cameras, their use in connection with optical traps will probably increase dramatically in the years to come. In order to obtain a position resolution with the imaging method that is comparable to the ~nm resolution, one may have with the nonimaging methods, though careful data analysis is required and the shape of the particle trapped and tracked must be well characterized. One may, for example, obtain subpixel resolution in particle tracking based on the calculation of the cross-correlation of a mask image with subsequent images of the tracked particle. Another concern with an imaging method is data storage and data transfer. On the other hand, with an imaging method, mechanical drift during an experiment may readily be subtracted from the data.

Optical-trapping interferometry was first successfully applied to investigate single-molecule motion with nanometer resolution (Svoboda et al., 1993). As the name indicates, optical trapping interferometry relies on interferometry between parts of the trapping light. Incoming trapping light of plane polarization is split by a Wollaston prism into physically separated beams of orthogonal polarization. After passing through the sample plane, the two traps are recombined in a second Wollaston prism. Subsequently, the polarization state of the recombined light is measured. This detection method is based on the same principle as DIC microscopy and thus requires the same optical components. Trapping interferometry is very sensitive, but only so along one dimension dictated by the Wollaston shear axis.

Another laser-based detection method makes use of a quadrant photodiode detector (QPD). By nature, it offers the possibility of making two-dimensional position detection through recording and subtracting signals measured from each of the four quadrants, just as in the detection of the vibrations of an atomic force microscope cantilever (Chapter 2). In addition, summation of the recorded currents from each of the four quadrants allows for detection of positions along the third dimension (Gittes and Schmidt, 1998). In the literature, two approaches to QPD detection has been demonstrated, either by detecting the trapping laser light in the back-focal plane of the optical trap (Allersma et al., 1998), or by detection of a different light source that follows the same optical path as the trapping laser (Veigel et al., 1998; Visscher and Block, 1998; Visscher et al., 1996). The latter requires alignment of two lasers, but the added complexity is outweighed by the opportunity for implementation of a feedback system with, for example, acousto-optic (Simmons et al., 1996; Visscher and Block, 1998) or electro-optic deflectors (Valentine et al., 2008). In addition, the detection laser may be chosen in the visible as long as it is very weak, thereby avoiding unintended filtering of above 1000 nm trapping laser light by common Si:PIN photodiodes (Berg-Sørensen et al., 2003). If direct detection of the trapping light of a wavelength around or above 1 μm is the most practical, good alternatives to the Si:PIN quadrant photodiode include diodes of a different material optimized for the task (Peterman et al., 2003b) or a position-sensitive detector (Neuman and Block, 2004).

4.7 FORCE CALIBRATION

As described previously, precise calculations of the forces exerted by the optical traps are quite involved and thus are not always available. In addition, the laser intensity in the trapping plane may not readily be known. Thus, traditionally, the force exerted by the trapping laser is found by calibration. In this selection, we discuss methods used in the literature to calibrate single-beam optical tweezers.

The conceptually simplest way to calibrate the force is through an escape force measurement, balancing the force from the optical trap with the Stokes drag. This method does not require an advanced position detection system, but can be performed with a simple CCD camera and an electronically controlled translation stage. Here, a bead of known radius r is trapped by the optical tweezers in a sample chamber containing a liquid of known viscosity η that is translated by a fixed velocity v. In practice, this is achieved by translating the sample chamber with a triangular periodic function. Without accounting for thermal fluctuating forces, the total force acting on the trapped bead is

$$\mathbf{F}_{tot} = \mathbf{F}_{OT} - \gamma_0 \mathbf{v}, \qquad (4.7)$$

where γ_0 is the Stokes drag coefficient $\gamma_0 = 6\pi\eta r$. The velocity of translation is now increased until the point where the bead is seen to leave the trap as a result of the friction from the surrounding liquid. The corresponding velocity is known as the escape velocity, v_{esc}, and thereby the maximum force that may be exerted by the optical trap, F_{esc}, is determined,

$$\mathbf{F}_{esc} = 6\pi\eta r \mathbf{v}_{esc} \qquad (4.8)$$

This type of calibration is useful when checking, for example, that the force varies linearly with the intensity of the laser.

However, a close inspection of the functional form of the optical trapping potential requires more involved experiments that would allow the experimenter to determine the relative position of the bead with respect to the minimum of the optical trap. This may, for example, be done by having two types of beads in the sample chamber within the field of view, one that is stuck to the chamber wall and one that is trapped. When the chamber starts moving, the difference in position of the trapped bead reflects its new equilibrium position, whereas imaging of beads that are stuck to the chamber walls allows for a determination of the actual velocity with which the chamber is moved. Let us, for simplicity, denote the position along the direction of the translation velocity v by x, in the rest frame of the optical trap and the camera used for recording. The force balance in Equation 4.7 states that when the trapped bead is in its (new) equilibrium position, $x_0(v)$, in the optical trap while the chamber is moved along x at a velocity v, the force from the optical trap may be determined as

$$F_{OT} = (x_0(v)) = \gamma_0 v \qquad (4.9)$$

A more precise method that also allows for closer inspection of the shape of the trapping potential is through measurements of the Brownian motion of a trapped bead. Also in this case, or at least in its simplest version, the bead radius and the viscosity of the surrounding liquid are assumed known. The methods allow, first, for the experimenter to check that the trapping potential is harmonic or, rather, within which range it stays harmonic. Then, by data analysis to be described in detail below, it supplies a value for the spring constant κ of the trapping potential. With a value for κ at hand, a precise measurement of the position of a trapped object relative to the potential minimum of the trap, x, will then allow for the value of the force, κx, to be known. For simplicity, we consider one dimension only. A similar analysis may be applied to independent, orthogonal coordinates y and z.

If the trapping potential is harmonic, the equation of motion of the trapped bead can be written as a Langevin equation,

$$m\ddot{x} = -\gamma_0 \dot{x} - \kappa x + F(t;T),$$ (4.10)

where $F(t)$ is a Langevin force accounting for the Brownian forces acting on the bead. This force depends parametrically on the temperature T as explicitly stated in Equation 4.10 but is omitted in what follows; this parametrical dependence appears in the correlation function

$$\langle F(t) \rangle = 0; \quad \langle F(t)F(t') \rangle = 2k_B T \gamma_0 \delta(t - t').$$ (4.11)

Equation 4.10 allows one to define the two timescales of the problem, an inertial timescale $t_{inertia} = m/\gamma_0$ describing the typical time for loss of kinetic energy to friction, and a timescale for relaxation of the bead in the trap, $t_{relax} = \gamma_0/\kappa$. If $t_{inertia} \ll t_{relax}$, corresponding to dynamics at low Reynolds number, the inertial term in Equation 4.10 can be neglected.

In the experiment, a bead is trapped and a time series of its positions, $x(t_i)$, $t_i = i\Delta t$, is recorded.

With N the number of positions recorded, we define the measurement time, $t_{meas} = N\Delta t$, and arrange for t_{meas} to be long compared to t_{relax}.

The functional form of the trapping potential is investigated via the thermal distribution of positions.

For a particle with potential energy described by a potential $U(x)$, the equilibrium distribution of positions is described according to

$$p(x) = A\exp\left(-\frac{U(x)}{k_B T}\right),$$ (4.12)

where A is an appropriate normalization constant. With $t_{meas} \gg t_{relax}$, we can use Equation 4.12 to check the functional form of the trapping potential, $U(x)$, and further determine the parameters of this trapping potential as we shall now describe.

The experimental time series of the position is converted into a distribution of positions visited by binning the x-axis in intervals of length Δx and counting the number of occurrences of a measurement $n(x_j)$ with $x \in [x_j - \Delta x/2, x_j + \Delta x/2]$ and $x_j - x_{j-1} = \Delta x$. These experimental data are described by the theoretical function

$$p^{(bin)}(j) \propto \int_{x_j - \Delta x/2}^{x_j + \Delta x/2} p(x)dx.$$ (4.13)

With the assumption of a harmonic trapping potential, we have

$$p(x) = \sqrt{\frac{2k_B T\pi}{\kappa}} \exp\left(-\frac{\kappa x^2}{k_B T}\right) \tag{4.14}$$

$$p^{(\text{bin})}(j) \propto \frac{k_B T\pi}{\kappa}\left[\text{erf}\left(\sqrt{\frac{\kappa}{2k_B T}}(x_j + \Delta x/2)\right) - \text{erf}\left(\sqrt{\frac{\kappa}{2k_B T}}(x_j - \Delta x/2)\right)\right] \tag{4.15}$$

and thus the binned position distribution is an error function rather than a Gaussian.

However, stochastic mechanical drift in the experimental setup will result in a widening of the histogram of positions visited. As a matter of fact, from looking at the raw time series, one may only with difficulty establish whether mechanical drift is present or not, as drift in most cases would result in a distribution of positions with Gaussian characteristics, just as that resulting from a harmonic trapping potential. All that requires is that the distribution of drift is normal, as is often the case due to the central limit theorem. In addition, if the position is recorded by a quadrant photodiode, an independent measurement of the conversion of voltage, the output from the photodiode, to a length unit is required.

Thus, it is natural to resort to other types of data analysis. Figure 4.6C illustrates two ways, either by the analysis of the power spectrum of the time series of positions or by analyzing the autocorrelation function, $C(\tau) \equiv \langle x(t)x(t+\tau)\rangle_t / \langle x(t)^2\rangle_t$, where the subscript t indicate averaging over time, t. If the inertial term on the left-hand side of Equation 4.10 is neglected and the stochastic nature of the Langevin force, Equation 4.11, is accounted for, one realizes that

$$C(\tau) = \exp(-\tau/t_{\text{relax}}). \tag{4.16}$$

Fitting this function to experimental data provides an evaluation of κ (Pralle et al, 1998).

A popular type of data analysis is through the analysis of the power spectrum from the time series of positions measured. The power spectrum is calculated from the Fourier transformed of the time series of positions,

$$\tilde{x}_k = \int_{-t_{\text{meas}}/2}^{t_{\text{meas}}/2} \exp(i2\pi f_k t)x(t)\,dt, \; f_k \equiv k/t_{\text{meas}}, \; k \text{ integer} \tag{4.17}$$

If we Fourier-transform the equation of motion, Equation 4.10, with uncorrelated stochastic Langevin forces, we find the expected experimental power spectrum, $P(f_k)$,

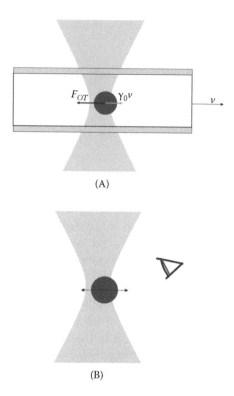

(A)

(B)

FIGURE 4.6 Illustration of different means to calibrate the optical tweezers. In (A), the Stokes drag procedure is illustrated, whereas (B) sketches the passive observation of Brownian motion of the bead in the trap. Panel C illustrates how the time series measured as in B may be analyzed: by an analysis of the histogram, by power spectral analysis, or by an analysis of the autocorrelation function of the positions.

$$P(f_k) \equiv \frac{1}{t_{meas}} \left\langle \left| \tilde{x}_k \right|^2 \right\rangle = \frac{D/(2\pi^2)}{f_c^2 + f_k^2} \tag{4.18}$$

where D is the diffusion coefficient, $D=k_BT/\gamma_0$, and f_c is the so-called *corner frequency*, $f_c=\kappa/(2\pi\gamma_0)=1/(2\pi t_{relax})$.

In practice, of course, the discrete Fourier transformed of the experimental data

$$\hat{x}_k = \Delta t \sum_{j=1}^{N} \exp(i2\pi f_k t_j) x(t_j) \tag{4.19}$$

is used instead of $\tilde{x}_k$.

Yet, as the finite sampling frequency in an actual experiment results in aliasing, the expression in Equation 4.18 does not describe the experimental data. Even with

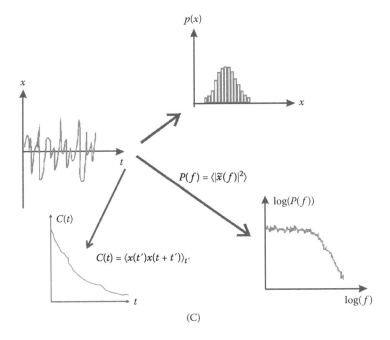

(C)

FIGURE 4.6 (Continued) Illustration of different means to calibrate the optical tweezers. In (A), the Stokes drag procedure is illustrated, whereas (B) sketches the passive observation of Brownian motion of the bead in the trap. Panel C illustrates how the time series measured as in B may be analyzed: by an analysis of the histogram, by power spectral analysis, or by an analysis of the autocorrelation function of the positions.

the use of antialiasing filters, aliasing should be accounted for—fortunately, this is easy in a computer program,

$$P^{(\text{aliased})}(f) = \sum_{n=-\infty}^{\infty} P^{(\text{theory})}\left(f + nf_{\text{sample}}\right) \qquad (4.20)$$

where $P^{(\text{theory})}$ is the function describing the theoretical model for the experimental situation, and $P^{(\text{aliased})}$ is the function to fit to the experimental spectrum recorded. In practice, the infinite sum may be approximated by a sum of $|n|$ below of order 10 to 15.

If antialiasing filters are in the setup, we recommend inclusion of the functional form for those in the theoretical function, $P^{(\text{theory})}$. For example, a first-order filter with a roll-off frequency $f_{3\text{dB}}$ reduces the power of its input by a factor

$$\frac{1}{1 + \left(f/f_{3\text{dB}}\right)^2}. \qquad (4.21)$$

In experimental setups where an Si:PIN photodiode is applied to detect infrared light of a wavelength above roughly 950 nm, unintended filtering appears (Berg-Sørensen et al., 2003; Peterman et al., 2003b) but may be accounted for by multiplication with an appropriate filter function, including one or two additional fitting parameters.

We are still not done: In a typical *in vitro* optical trapping experiment, the trapped bead is close to a surface whereas the expression for the Stokes drag, $\gamma_0 = 6\pi\eta r$, corresponds to flow far from any surface.

Apart from the dependence of distance to a surface expressed through Faxén's law, the frictional force also depends on frequency. Both of these corrections are accounted for by the expression

$$P^{(\text{hydro})}(f;R/\ell) = \frac{D/(2\pi^2)\dfrac{\text{Re}\,\gamma}{\gamma_0}}{\left(f_c + f\dfrac{\text{Im}\,\gamma}{\gamma_0} - f^2/f_m\right)^2 + \left(f\dfrac{\text{Re}\,\gamma}{\gamma_0}\right)^2}$$

(4.22)

where

$$\text{Re}\,\gamma/\gamma_0 = 1 + \sqrt{f/f_v} - \frac{3R}{16\ell} + \frac{3R}{4\ell}\exp\left(-\frac{2\ell}{R}\sqrt{f/f_v}\right)\cos\left(\frac{2\ell}{R}\sqrt{f/f_v}\right)$$

and

$$\text{Im}\,\gamma/\gamma_0 = -\sqrt{f/f_v} + \frac{3}{4}\frac{R}{\ell}\exp\left(-\frac{2\ell}{R}\sqrt{f/f_v}\right)\sin\left(\frac{2\ell}{R}\sqrt{f/f_v}\right).$$

In these expressions, two new characteristic frequencies appear: $f_v \equiv \nu/(\pi R^2)$ and $f_m \equiv \gamma_0/(2\pi m^*)$ with ν the kinematic viscosity of the liquid, $\nu = \eta/\rho$ and $m^* \equiv m + 2\pi\rho R^3/3$. The density ρ is that of the liquid.

Figure 4.7 illustrates a fit to experimental data, striving for high precision. The thin solid line demonstrates the Lorentzian theory, Equation 4.18, but accounting for aliasing, Equation 4.20. The dashed line accounts for the distance and frequency-dependent hydrodynamic corrections, Equation 4.22. The data are recorded by an Si:PIN QPD in the back focal plane of the condenser, and the trapping laser has a wavelength of 1064 nm. Therefore, unintended filtering must be accounted for, as demonstrated by the perfect fit with the thick solid line. A more detailed theoretical description of the procedure may be found in the work of Berg-Sørensen and Flyvbjerg (2004), and MATLAB® codes are available from *Computer Physics Communications* (Hansen et al., 2006; Tolic-Nørrelykke et al., 2004a). Also, holographic optical traps have successfully been calibrated by the power spectral analysis just described (Van Der Horst and Forde, 2008).

With calibration of the trap based on a passive observation of the Brownian motion of a bead in the trap, the conversion of voltage recorded by the QPD to true physical length unit requires either an independent calibration of the latter or is done by considering D in, for example, Equation 4.22 as a fitting parameter. However, this purely passive observation may be supplemented by a procedure in which the

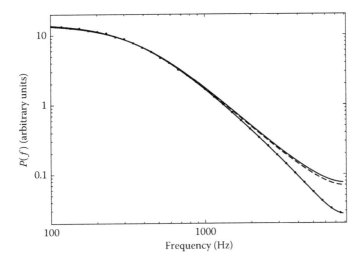

FIGURE 4.7 Illustration of a detailed power spectral analysis, accounting for hydrody-namic corrections as well as unintended filtering by the photodetection system, as described in (Berg-Sørensen, K. and Flyvbjerg, H. *Review of Scientific Instruments,* 75, 594–612, 2004). The thick solid line indicates the fit of the full theoretical expression to the data points (black dots). The thin solid line shows the Lorentzian theory, Equation 4.18, but accounting for aliasing, and the dashed line shows the hydrodynamically correct theory, Equation 4.22, with aliasing accounted for. The free parameters have been chosen equal to those resulting from the fit of the full theoretical expression to the experimental data. In this full theoretical expression, unintended filtering by the photodiode detection system is accounted for too.

trap is oscillated at a fixed frequency, f_0 (Tolic-Nørrelykke et al., 2006; Vermeulen et al., 2006). In this case, the power spectrum is composed of a sum of the power spectrum in Equation 4.22 and a δ-function term at the frequency f_0. The amplitude of the power spectral peak at f_0 readily gives information on the conversion between voltage recorded at the QPD and the excursion in the physical unit, length.

In cases where a dielectric microsphere or a metallic nanosphere is trapped in a vis-coelastic medium rather than a simple viscous fluid, a combination of measurements of both passive and active nature has been suggested (Fischer and Berg-Sørensen, 2007; Fischer et al., 2010). Under restricted conditions for the active measurements, this combination of data series allows for both investigations of trap characteristics and of viscoelastic properties of the medium. The procedure involves a series of experiments, as illustrated in Figure 4.8. In passive measurements, time series of posi-tions are recorded and power spectra are calculated. In active measurements, oscillat-ing either the sample stage or the trapping laser, time series of positions of the trapped bead are recorded and compared to the position of the oscillating trap or stage, and a so-called relaxation spectrum is recorded. Subsequently, the data from active and passive spectral measurements are combined to extract the information sought for.

It has also been suggested to use the limiting behavior of the real part of the viscoelastic modulus, $\lim_{\omega \to 0} G'(\omega)$ for the determination of the spring constant of the trap in a viscoelastic medium (Atakhorrami et al., 2006). Another alternative that

(A) Force calibration (FDT method), passive part

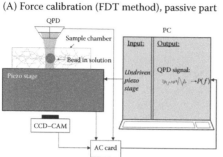

(B) Force calibration (FDT method), active part

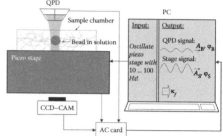

(C) Direct positional calibration (sinusoidal stage driving)

(D) Pixel size calibration (displacement of stage by a known distance)

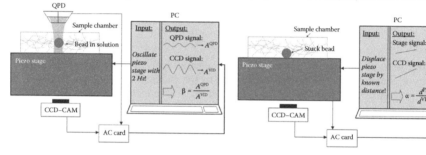

(E) Phase correction calibration for calibrating delay between AC card channels (sinusoidal stage driving)

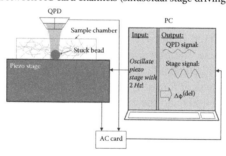

FIGURE 4.8 Illustration of the suggested method for calibration of the optical trap while, at the same time, determining the viscoelastic properties of the medium in which the probe is trapped. (Reprinted from Fischer, M., Richardson, A. C., Reihani, S. N. S., Oddershede, L. B., and Berg-Sørensen, K. *Review of Scientific Instruments*, 81, 2010. With permission. Copyright 2010, American Institute of Physics, where the need for and the details of each step (A) to (E) are explained.)

may be used when the object to trap is well known and available both in the viscoelastic medium and in a normal viscous liquid is to simply determine the spring constant of the trap in the normal viscous liquid through one of the methods just described. Then, by proper accounting for the difference in refractive index between the two media, the spring constant in the viscoelastic one may be found.

These methods are suggested for quantitative force measurements in *in vivo* systems where also active processes, such as action of molecular motors, may take place. These active processes complicate the analysis, which is typically based on the assumption that concepts of equilibrium thermodynamics are applicable.

A completely different methodology determines the force exerted by the optical trap through detection of scattered light (Smith et al., 1996, 2003). However, this requires the collection of all scattered photons.

4.8 *IN VITRO* APPLICATIONS

Optical traps have been applied extensively to investigate single-molecule systems, reconstructed to function *in vitro*. Here, we will review a number of examples that have demonstrated the applicability of optical traps while at the same time requesting technical improvements to the setup or a different experimental assay.

4.8.1 MOLECULAR MOTORS

Optical trapping with single-molecule position resolution was first demonstrated in investigations of the molecular motor kinesin (Svoboda et al., 1993), where 8 nm steps of kinesin forward motion on their substrate, microtubules, were recorded. A single kinesin molecule was bound to a polystyrene microsphere and trapped by the single-beam optical tweezers. By the addition of a feedback system based on acousto-optic deflectors, the setup was developed to ensure that the optical trap exerted a constant force on the kinesin molecule. With this force clamp, measurement statistics was improved to confirm that one 8 nm step forward was powered by the hydrolysis of one molecule of ATP (Schnitzer and Block, 1997) and that the kinetics can be described similarly to the standard Michaelis–Menten enzyme kinetics (Visscher et al., 1999).

In this assay, the substrate, microtubules, is attached to a glass coverslide, and the transport of the kinesin molecule is observed. These measurements take place close to a surface, and it is therefore important to account for the actual distance to the surface when analyzing the data as the frictional force depends on distance to the coverslide.

Another molecular motor first studied with a similar assay is RNA polymerase (RNAp). In that case, the substrate is the double-stranded DNA, one end of which is attached to a glass coverslide, whereas the other is being processed by the RNAp. The RNAp is coupled to a microsphere held in the optical trap (Wang et al., 1998). In the quest for an understanding of the chemomechanics of the transcription, some measurements indicated load-dependent pauses in the transcription (Forde et al., 2002). Some of the issues under debate were resolved with an experimental assay that allows for base-pair position resolution (Abbondanzieri et al., 2005). It is based on a dual-beam optical tweezers system, as sketched in Figure 4.9A. In order to reach both the high resolution and high bandwidth required for base-pair resolution in the RNAp–DNA system, this dual-beam optical tweezers assay does not involve a feedback system but is based on an all-passive optical force clamp (Greenleaf et al., 2005). One of the optical traps is weak, whereas the other is strong. The bead in the weak trap is displaced in the trap to the point where the trapping potential is anharmonic, that is, where the potential flattens and the trapping force essentially vanishes.

With this assay, the sample is no longer close to a surface and, by proper engineering of the attachment between dielectric beads and biomolecules, their orientation may be fully controlled.

Molecular motors may either be processive or nonprocessive. For investigations of nonprocessive motors, such as myosin-I interacting with actin filaments, a three-bead, dual-beam optical tweezers assay is used (Finer et al., 1994). This assay is illustrated in Figure 4.9B. Essentially, the same assay has also proved useful for the investigation of processive molecular motors, such as myosin-V (Veigel et al., 2002). When a molecular motor from the untrapped, larger bead attaches to the actin filament, held taut between the two optical traps; a change in position of one of the beads on the actin filament is observed. Investigations of both the change in position and the reduced thermal noise in the signal provide detailed information on the molecular motor.

A different type of molecular motor for which investigations applying an optical tweezers assay has also proved successful is the packaging motor of the virus φ89. In this case, the force required to package the viral DNA inside the viral capsid has been measured in an assay, where a single optical tweezer is combined with a micropipette. The assay is illustrated in Figure 4.9C, for a different type of biological problem. When investigating φ89 with this essay, one end of the DNA is coupled to a microsphere and the viral capsid is as well. Then, one bead is controlled by the micropipette and the second by the optical tweezers, which in addition provides the ability to measure forces. In principle, the same type of assay may be applied for investigations of the RNAp–DNA system, as illustrated in the figure.

4.8.2 OTHER SYSTEMS

Apart from investigations of the single-molecule dynamics of molecular motors, first applications of optical tweezers include experiments where the elasticity of DNA was investigated (Smith et al., 1996), in the assay where a single-beam optical tweezers is combined with a micropipette. A novel model for the elasticity of entropic springs (Marko and Siggia, 1995) was developed, resulting in the now-famous Marko–Siggia interpolation formula for the elasticity of wormlike chains (WLC), which was demonstrated to describe the experimental data very well (Bustamante et al., 1994). Also, the folding of the giant muscle protein, titin, was investigated in a similar experimental setup (Kellermayer et al., 1997), and interpreted through a combination of WLC-like stretching and unfolding of domains in the protein.

Measurement of the forces involved in unzipping of DNA has also been demonstrated (Bockelmann et al., 2002), in an assay based on the same principles as the optical trapping interferometer assay used for the first investigations of kinesin (Svoboda et al., 1993).

The dual-beam optical tweezers assay and the assay of single-beam optical tweezers in combination with a micropipette (Figure 4.9A,C) have also been applied in investigations of folding and unfolding of RNA hairpins (Liphardt et al., 2002; Woodside et al., 2006). The scrutiny of the folding landscapes for the RNA relies on measurements in systems that are not in thermodynamic equilibrium, yet, through

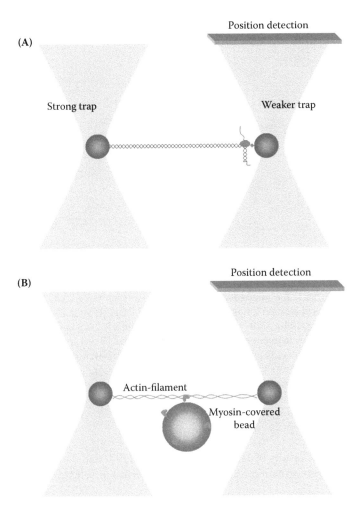

FIGURE 4.9 (A) Optical tweezers assay with two traps, a strong and a weaker one. This setup has been applied to determine the processing of RNA polymerase with a spatial resolution sufficiently good to resolve the processing of individual base pairs (Abbondanzieri, E. A., Greenleaf, W. J., Shaevitz, J. W., Landick, R, and Block, S. M. *Nature,* 438, 460–465, 2005). (B) Three-bead optical tweezers assays, with two traps. This assay was originally developed for the investigation of nonprocessive molecular motors (myosin-I) (Finer, J. T., Simmons, R. M., and Spudich, J. A. *Nature,* 368, 113–119, 1994), but has later been used also for the study of processive molecular motors, such as myosin-V (Veigel, C., Wang, F., Bartoo, M. L., Sellers, J. R., and Molloy, J. E. *Nature Cell Biology,* 4, 59–65, 2002). When a molecular motor from the untrapped, larger bead, attaches to the actin filament, held between smaller beads trapped in the two optical traps, a change in position of one of the smaller beads on the actin filament is observed. By investigations of both the change in position and the reduced thermal noise in the signal, detailed information about the molecular motor investigated has been found.

(C)

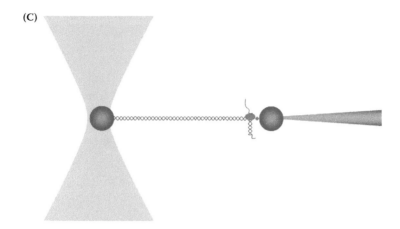

FIGURE 4.9 (CONTINUED)   (C) Optical tweezers assay with one optical trap, combined with a micropipette. The fundamentals of this assay was originally developed and applied to investigate the stretching of DNA (Smith, S. B., Cui, Y. J., and Bustamante, C. *Science,* 271, 795–799, 1996), but if the micropipette is rotated rather than pulled, an experiment may be designed that investigates the torque of double-stranded DNA (Bryant, Z., Stone, M. D., Gore, J., Smith, S. B., Cozzarelli, N. R., and Bustamante, C. *Nature,* 424, 338–341, 2003). Other versions of the basic assay (a micropipette and an optical trap) have been applied to study the portal motor of the phage φ89 (Chemla, Y. R., Aathavan, K., Michaelis, J., Grimes, S., Jardine, P. J., Anderson, D. L., and Bustamante, C. *Cell,* 122, 683–692, 2005), the unfolding of the muscle protein, titin (Kellermayer, M. S. Z., Smith, S. B., Granzier, H. L., and Bustamante, C. *Science,* 276, 1112–1116, 1997), and the unfolding of RNA hairpins, to test new theories for nonequilibrium thermodynamics (Liphardt, J., Dumont, S., Smith, S. B., Tinoco Jr, I., and Bustamante, C. *Science,* 296, 1832–1835, 2002).

the application of results of Jarzynski (Jarzynski, 1997) and Crooks (Crooks, 1998, 1999), allow for information on equilibrium properties to be extracted.

4.9 *IN VIVO* APPLICATIONS

Already from the early development of optical trapping, *in vivo* systems were investigated, as in the early trapping of bacteria and viruses (Ashkin et al., 1986; Ashkin et al., 1987), in investigations of the locomotion of sperm cells (Konig et al., 1996), growth of neuronal growth cones (Dai and Sheetz, 1995), motility of keratocytes (Galbraith and Sheetz, 1999), or forces exerted by different types of bacterial motors (Berry and Berg, 1997; Maier et al., 2002).

In recent years, as optical tweezers have transformed from a tool of manipulation to a tool of precise force spectroscopy for *in vitro* systems, the same technology has been sought applied in *in vivo* investigations focused on precise measurements of intracellular processes. One question to address is that of measurements of the viscoelastic properties of the cytoplasm of living cells (Tolic-Nørrelykke et al., 2004b), closely linked to microrheological measurements on reconstituted cytoskeletal protein networks where laser tracking with very weak single- or dual-beam

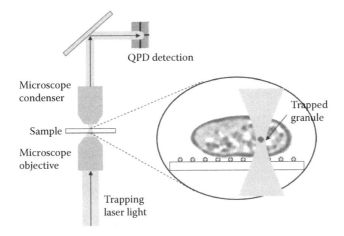

QPD detection

Microscope
condenser

Trapped
granule

Sample

Microscope
objective

Trapping
laser light

FIGURE 4.10 Figure showing the optical tweezers assay to trap and manipulate naturally occurring granuli in living *S. pombe* cells. Trapping of the lipid granuli has been applied both in experiments where the cytoplasm has been investigated as a viscoelastic medium (Selhuber-Unkel, C., Yde, P., Berg-Sørensen, K., and Oddershede, L. B. *Physical Biology,* 6, 2009; Tolic-Nørrelykke, I. M., Munteanu, E. L., Thon, G., Oddershede, L., and Berg-Sørensen, K. *Physical Review Letters,* 93, 078102-1–078102-4, 2004b) and in experiments where the cell nucleus has been displaced (Tolic-Nørrelykke, I. M., Sacconi, L., Stringari, C., Raabe, I., and Pavone, F. S. *Current Biology,* 15, 1212–1216, 2005).

optical tweezers has proved very useful. In studies of the yeast cell *S. pombe*, naturally occurring lipid granuli were used as tracer particles for investigations of the surrounding medium, as illustrated in Figure 4.10.

Single-beam optical tweezers have also been applied to move intracellular organelles, in a combination of optical trapping in the infrared (970 nm) with laser ablation by an intense-pulsed laser at a lower wavelength of 895 nm. In these experiments, a lipid granulus was trapped and moved in a raster-scanning manner close to the cell nucleus, resulting in a translation of the nucleus relative to the cell center, as illustrated in Figure 4.11. This type of experiment has been very instructive in elucidating the role of polymerizing and depolymerizing microtubules in the positioning of the cell nucleus (Tolic-Nørrelykke et al., 2005).

4.10   FUTURE DIRECTIONS

As has been discussed in this chapter, optical trapping has within the last decade developed from being a "simple" tool of manipulation to a very precise force scope, at least in *in vitro* applications. In the years to come, we foresee the continued merger of methods of optical trapping with other methods applied in single-molecule biophysics, such as imaging through single-molecule fluorescence. Simultaneous and co-localized trapping and fluorescent detection have been demonstrated in studies of unzipping of DNA (Lang et al., 2004) in an *in vitro* setup. However, *in vivo* experiments have also been performed that demonstrate the applicability of these methods

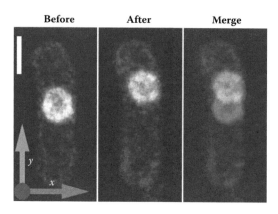

FIGURE 4.11   (See color insert.) Figure demonstrating the displacement of the cell nucleus in living *S. pombe* cells. Imaging is enhanced by fluorescence microscopy. The scale bar in the left corner is 3 μm long. Lipid granuli were trapped and moved in a raster-scanning manner to exert a force on the cell nucleus. (Reproduced from Maghelli, N. and Tolic-Nørrelykke, I. M. *Journal of Biophotonics*, Vol. 1, pp 299–309, 2008. Copyright Wiley-VCH Verlag GmbH & Co. KGaA. With permission.)

(Sims and Xie, 2009). In this particular example, the motility of presumably single motor proteins transporting endogenous lipid droplets under load was observed. In addition to this merger of single-molecule methods, new biological questions that request better resolution or better precision will continue to push the optical trapping techniques toward higher spatial and/or temporal resolution.

REFERENCES

Abbondanzieri, E. A., Greenleaf, W. J., Shaevitz, J. W., Landick, R., and Block, S. M. (2005). Direct observation of base-pair stepping by RNA polymerase. *Nature*, 438, 460–465.

Allersma, M. W., Gittes, F., Decastro, M. J., Stewart, R. J., and Schmidt, C. F. (1998). Two-dimensional tracking of ncd motility by back focal plane interferometry. *Biophysical Journal*, 74, 1074–1085.

Ashkin, A. (1970). Acceleration and trapping of particles by radiation pressure. *Physical Review Letters*, 24, 156–159.

Ashkin, A. (1997). Optical trapping and manipulation of neutral particles using lasers. *Proceedings of the National Academy of Sciences of the United States of America*, 94, 4853–4860.

Ashkin, A. and Dziedzic, J. M. (1987). Optical trapping and manipulation of viruses and bacteria. *Science*, 235, 1517–1520.

Ashkin, A., Dziedzic, J. M., Bjorkholm, J. E., and Chu, S. (1986). Observation of a single-beam gradient force optical trap for dielectric particles. *Optics Letters*, 11, 288–290.

Ashkin, A., Dziedzic, J. M., and Yamane, T. (1987). Optical trapping and manipulation of single cells using infrared laser beams. *Nature*, 330, 769–771.

Atakhorrami, M., Sulkowska, J. I., Addas, K. M., Koenderink, G. H., Tang, J. X., Levine, A. J., Mackintosh, F. C., and Schmidt, C. F. (2006). Correlated fluctuations of microparticles in viscoelastic solutions: Quantitative measurement of material properties by microrheology in the presence of optical traps. *Physical Review E*, 73. Art no. 021875.

Berg-Sørensen, K. and Flyvbjerg, H. (2004). Power spectrum analysis for optical tweezers. *Review of Scientific Instruments*, 75, 594–612.

Berg-Sørensen, K., Oddershede, L., Florin, E. L., and Flyvbjerg, H. (2003). Unintended filtering in a typical photodiode detection system for optical tweezers. *Journal of Applied Physics*, 93, 3167–3176.

Berry, R. M. and Berg, H. C. (1997). Absence of a barrier to backwards rotation of the bacterial flagellar motor demonstrated with optical tweezers. *Proceedings of the National Academy of Sciences of the United States of America*, 94, 14433–14437.

Bockelmann, U., Thomen, P., Essevaz-Roulet, B., Viasnoff, V., and Heslot, F. (2002). Unzipping DNA with optical tweezers: High sequence sensitivity and force flips. *Biophysical Journal*, 82, 1537–1553.

Bryant, Z., Stone, M. D., Gore, J., Smith, S. B., Cozzarelli, N. R., and Bustamante, C. (2003). Structural transitions and elasticity from torque measurements on DNA. *Nature*, 424, 338–341.

Burns, M. M., Fournier, J. M., and Golovchenko, J. A. (1989). Optical binding. *Physical Review Letters*, 63, 1233–1236.

Bustamante, C., Marko, J. F., Siggia, E. D., and Smith, S. (1994). Entropic elasticity of λ-phage DNA. *Science*, 265, 1599–1600.

Chemla, Y. R., Aathavan, K., Michaelis, J., Grimes, S., Jardine, P. J., Anderson, D. L., and Bustamante, C. (2005). Mechanism of force generation of a viral DNA packaging motor. *Cell*, 122, 683–692.

Chu, S. (1998). The manipulation of neutral particles. *Reviews of Modern Physics*, 70, 685–706.

Crooks, G. E. (1999). Entropy production fluctuation theorem and the nonequilibrium work relation for free energy differences. *Physical Review E—Statistical Physics, Plasmas, Fluids, and Related Interdisciplinary Topics*, 60, 2721–2726.

Crooks, G. E. (1998). Nonequilibrium measurements of free energy differences for microscopically reversible Markovian systems. *Journal of Statistical Physics*, 90, 1481–1487.

Dai, J. and Sheetz, M. P. (1995). Mechanical properties of neuronal growth cone membranes studied by tether formation with laser optical tweezers. *Biophysical Journal*, 68, 988–996.

Dholakia, K. and Lee, W. M. (2008). Optical trapping takes shape: The use of structured light fields. *Advances in Atomic, Molecular and Optical Physics*, 56, 261–337.

Dienerowitz, M., Mazilu, M., and Dholakia, K. (2008). Optical manipulation of nanoparticles: A review. *Journal of Nanophotonics*, 2. Art no. 021875

Eriksen, R. L., Mogensen, P. C., and Gluckstad, J. (2002). Multiple-beam optical tweezers generated by the generalized phase-contrast method. *Optics Letters*, 27, 267–269.

Finer, J. T., Simmons, R. M., and Spudich, J. A. (1994). Single myosin molecule mechanics: Piconewton forces and nanometre steps. *Nature*, 368, 113–119.

Fischer, M. and Berg-Sørensen, K. (2007). Calibration of trapping force and response function of optical tweezers in viscoelastic media. *Journal of Optics A-Pure and Applied Optics*, 9, S239–S250.

Fischer, M., Richardson, A. C., Reihani, S. N. S., Oddershede, L. B., and Berg-Sørensen, K. (2010). Active-passive calibration of optical tweezers in viscoelastic media. *Review of Scientific Instruments*, 81. Art no. 015103.

Forde, N. R., Izhaky, D., Woodcock, G. R., Wuite, G. J. L., and Bustamante, C. (2002). Using mechanical force to probe the mechanism of pausing and arrest during continuous elongation by Escherichia coli RNA polymerase. *Proceedings of the National Academy of Sciences of the United States of America*, 99, 11682–11687.

Galbraith, C. G. and Sheetz, M. P. (1999). Keratocytes pull with similar forces on their dorsal and ventral surfaces. *Journal of Cell Biology*, 147, 1313–1323.

Gittes, F. and Schmidt, C. F. (1998). Interference model for back-focal-plane displacement detection in optical tweezers. *Optics Letters, 23,* 7–9.

Greenleaf, W. J., Woodside, M. T., Abbondanzieri, E. A., and Block, S. M. (2005). Passive all-optical force clamp for high-resolution laser trapping. *Physical Review Letters, 95.* Art no. 208102.

Grier, D. G. (2003). A revolution in optical manipulation. *Nature, 424,* 810–816.

Guck, J., Ananthakrishnan, R., Mahmood, H., Moon, T. J., Cunningham, C. C., and Kas, J. (2001). The optical stretcher: A novel laser tool to micromanipulate cells. *Biophysical Journal, 81,* 767–784.

Guck, J., Ananthakrishnan, R., Moon, T. J., Cunningham, C. C., and Käs, J. (2000). Optical deformability of soft biological dielectrics. *Physical Review Letters, 84,* 5451–5454.

Hansen, P. M. and Oddershede, L. B. (2005). Optical trapping inside living organisms. *Proceedings of SPIE—The International Society for Optical Engineering.*

Hansen, P. M., Tolic-Nørrelykke, I. M., Flyvbjerg, H., and Berg-Sørensen, K. (2006). Tweezercalib 2.1: Faster version of MatLab package for precise calibration of optical tweezers. *Computer Physics Communications, 175,* 572–573.

Huang, Y. S., Yeh, C. L., Liao, G. B., Chen, Y. F., Tsai, T. F., and Chiou, A. (2008). Deformability of mice erythrocytes measured by oscillatory optical tweezers. *Proceedings of SPIE— The International Society for Optical Engineering.*

Jarzynski, C. (1997). Nonequilibrium equality for free energy differences. *Physical Review Letters, 78,* 2690–2693.

Jauffred, L., Richardson, A. C., and Oddershede, L. B. (2008). Three-Dimensional optical control of individual quantum dots. *Nano Letters, 8,* 3376–3380.

Kellermayer, M. S. Z., Smith, S. B., Granzier, H. L., and Bustamante, C. (1997). Folding-unfolding transitions in single titin molecules characterized with laser tweezers. *Science,* 276, 1112–1116.

Konig, K., Svaasand, L., Liu, Y., Sonek, G., Patrizio, P., Tadir, Y., Berns, M. W., and Tromberg, B. J. (1996). Determination of motility forces of human spermatozoa using an 800 nm optical trap. *Cellular and Molecular Biology (Noisy-le-Grand, France),* 42, 501–509.

Lang, M. J., Fordyce, P. M., Engh, A. M., Neuman, K. C., and Block, S. M. (2004). Simultaneous, coincident optical trapping and single-molecule fluorescence. *Nature Methods,* 1, 133–139.

Lee, W. M., Reece, P. J., Marchington, R. F., Metzger, N. K., and Dholakia, K. (2007). Construction and calibration of an optical trap on a fluorescence optical microscope. *Nature Protocols,* 2, 3226–3238.

Lincoln, B., Schinkinger, S., Travis, K., Wottawah, F., Ebert, S., Sauer, F., and Guck, J. (2007). Reconfigurable microfluidic integration of a dual-beam laser trap with biomedical applications. *Biomedical Microdevices,* 9, 703–710.

Liphardt, J., Dumont, S., Smith, S. B., Tinoco, I., Jr, and Bustamante, C. (2002). Equilibrium information from nonequilibrium measurements in an experimental test of Jarzynski's equality. *Science,* 296, 1832–1835.

Maghelli, N. and Tolic-Nørrelykke, I. M. (2008). Versatile laser-based cell manipulator. *Journal of Biophotonics,* 1, 299–309.

Maier, B., Potter, L., So, M., Seifert, H. S., and Sheetz, M. P. (2002). Single pilus motor forces exceed 100 pN. *Proceedings of the National Academy of Sciences of the United States of America,* 99, 16012–16017.

Marko, J.F. and Siggia, E.D. (1995). Stretching DNA. *Macromolecules,* 28, 8759–8770.

Neuman, K. C. and Block, S. M. (2004). Optical trapping. *Review of Scientific Instruments,* 75, 2787–2809.

Neuman, K. C., Chadd, E. H., Liou, G. F., Bergman, K., and Block, S. M. (1999). Characterization of photodamage to Escherichia coli in optical traps. *Biophysical Journal,* 77, 2856–2863.

Pralle, A., Florin, E., L., Stelzer, E.H.K., and Hörber, J.K.H. (1998). Local viscosity probed by photonic force microscopy. *Applied Physics A*, 66, 571–573.

Peterman, E. J. G., Gittes, F., and Schmidt, C. F. (2003a). Laser-induced heating in optical traps. *Biophysical Journal*, 84, 1308–1316.

Peterman, E. J. G., Van Dijk, M. A., Kapitein, L. C., and Schmidt, C. F. (2003b). Extending the bandwidth of optical-tweezers interferometry. *Review of Scientific Instruments*, 74, 3246–3249.

Reicherter, M., Haist, T., Wagemann, E. U., and Tiziani, H. J. (1999). Optical particle trapping with computer-generated holograms written on a liquid-crystal display. *Optics Letters*, 24, 608–610.

Reihani, S. N. S. and Oddershede, L. B. (2007). Optimizing immersion media refractive index improves optical trapping by compensating spherical aberrations. *Optics Letters*, 32, 1998–2000.

Richardson, A. C., Reihani, S. N. S., and Oddershede, L. B. (2008). Non-harmonic potential of a single beam optical trap. *Optics Express*, 16, 15709–15717.

Rohrbach, A. (2005). Stiffness of optical traps: Quantitative agreement between experiment and electromagnetic theory. *Physical Review Letters*, 95, 1–4.

Roosen, G. (1977). A theoretical and experimental study of the stable equilibrium positions of spheres levitated by two horizontal laser beams. *Optics Communications*, 21, 189–194.

Sacconi, L., Tølic-Norrelykke, I. M., Stringari, C., Antolini, R., and Pavone, F. S. (2005). Optical micromanipulations inside yeast cells. *Applied Optics*, 44, 2001–2007.

Schnitzer, M. J. and Block, S. M. (1997). Kinesin hydrolyses one Atp per 8-nm step. *Nature*, 388, 386–390.

Selhuber-Unkel, C., Yde, P., Berg-Sørensen, K., and Oddershede, L. B. (2009). Variety in intracellular diffusion during the cell cycle. *Physical Biology*, 6. Art no. 025015

Simmons, R. M., Finer, J. T., Chu, S., and Spudich, J. A. (1996). Quantitative measurements of force and displacement using an optical trap. *Biophysical Journal*, 70, 1813–1822.

Sims, P. A. and Xie, X. S. (2009). Probing dynein and kinesin stepping with mechanical manipulation in a living cell. *ChemPhysChem*, 10, 1511–1516.

Smith, S. B., Cui, Y. J., and Bustamante, C. (1996). Overstretching B-DNA: The elastic response of individual double-stranded and single-stranded DNA molecules. *Science*, 271, 795–799.

Smith, S. B., Cui, Y. J., and Bustamante, C. (2003). Optical-trap force transducer that operates by direct measurement of light momentum. *Biophotonics, Pt B*, 361, 134–162.

Svoboda, K., Schmidt, C. F., Schnapp, B. J., and Block, S. M. (1993). Direct observation of kinesin stepping by optical trapping interferometry. *Nature*, 365, 721–727.

Tolic-Nørrelykke, I. M., Berg-Sørensen, K., and Flyvbjerg, H. (2004a). MatLab program for precision calibration of optical tweezers. *Computer Physics Communications*, 159, 225–240.

Tolic-Nørrelykke, I. M., Munteanu, E. L., Thon, G., Oddershede, L., and Berg-Sørensen, K. (2004b). Anomalous diffusion in living yeast cells. *Physical Review Letters*, 93, 078102-1–078102-4.

Tolic-Nørrelykke, I. M., Sacconi, L., Stringari, C., Raabe, I., and Pavone, F. S. (2005). Nuclear and division-plane positioning revealed by optical micromanipulation. *Current Biology*, 15, 1212–1216.

Tolic-Nørrelykke, S. F., Schaffer, E., Howard, J., Pavone, F. S., Julicher, F., and Flyvbjerg, H. (2006). Calibration of optical tweezers with positional detection in the back focal plane. *Review of Scientific Instruments*, 77. Art no. 013704

Valentine, M. T., Guydosh, N. R., Gutierrez-Medina, B., Fehr, A. N., Andreasson, J. O., and Block, S. M. (2008). Precision steering of an optical trap by electro-optic deflection. *Optics Letters*, 33, 599–601.

Van Der Horst, A. and Forde, N. R. (2008). Calibration of dynamic holographic optical tweezers for force measurements on biomaterials. *Optics Express,* 16, 20987–21003.

Veigel, C., Bartoo, M. L., White, D. C. S., Sparrow, J. C., and Molloy, J. E. (1998). The stiffness of rabbit skeletal actomyosin cross-bridges determined with an optical tweezers transducer. *Biophysical Journal,* 75, 1424–1438.

Veigel, C., Wang, F., Bartoo, M. L., Sellers, J. R., and Molloy, J. E. (2002). The gated gait of the processive molecular motor, myosin V. *Nature Cell Biology,* 4, 59–65.

Vermeulen, K. C., Van Mameren, J., Stienen, G. J. M., Peterman, E. J. G., Wuite, G. J. L., and Schmidt, C. F. (2006). Calibrating bead displacements in optical tweezers using acousto-optic deflectors. *Review of Scientific Instruments,* 77. Art no. 013704.

Visscher, K. and Block, S. M. (1998). Versatile optical traps with feedback control. *Methods in Enzymology.* vol. 298, 460–489.

Visscher, K., Gross, S. P., and Block, S. M. (1996). Construction of multiple-beam optical traps with nanometer-resolution position sensing. *IEEE Journal on Selected Topics in Quantum Electronics,* 2, 1066–1076. 460–489.

Visscher, K., Schnitzer, M. J., and Block, S. M. (1999). Single kinesin molecules studied with a molecular force clamp. *Nature,* 400, 184–189.

Wang, M. D., Schnitzer, M. J., Yin, H., Landick, R., Gelles, J., and Block, S. M. (1998). Force and velocity measured for single molecules of RNA polymerase. *Science,* 282, 902–907.

Woodside, M. T., Behnke-Parks, W. M., Larizadeh, K., Travers, K., Herschlag, D., and Block, S. M. (2006). Nanomechanical measurements of the sequence-dependent folding landscapes of single nucleic acid hairpins. *Proceedings of the National Academy of Sciences of the United States of America,* 103, 6190–6195.

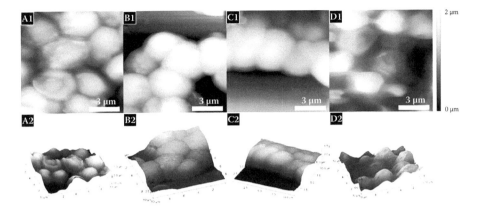

FIGURE 2.17 Tapping-mode AFM images of yeast cells in water using (A) mechanical actuation and optical readout, (B) mechanical actuation and piezoresistive readout, (C) thermal actuation and optical readout, and (D) thermal actuation and piezoresistive readout. (Reproduced with permission from IOP Publishing from Fantner, G. E., Schumann, W., Barbero, R. J., Deutschinger, A., Todorov, V., Gray, D. S., Belcher, A. M., Rangelow, I. W., and Youcef-Toumi, K. *Nanotechnology,* 20, 10, 2009.)

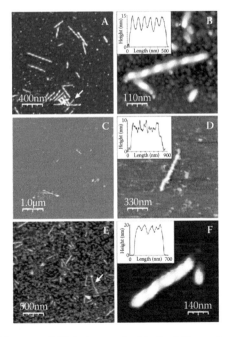

FIGURE 2.19 AFM images of Alzheimer paired helical filaments using DM AFM in air (A & B), DM AFM in liquid (C & D), and JM AFM in liquid (E & F). Reprinted from Moreno-Herrero, F., Colchero, J., Gomez-Herrero, J., Baro, A. M., and Avila, J. *European Polymer Journal,* 40, 927–932, 2004. Copyright 2004. With permission from Elsevier.)

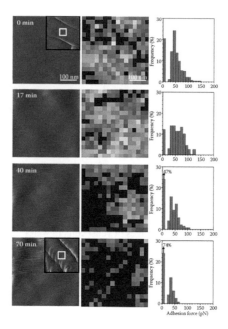

FIGURE 2.22 Real-time AFM imaging of the Mycobacterium surface after the introduction of pullulanase enzyme, showing the slow change in both surface roughness and decrease in adhesion events. (Reproduced from Verbelen, C. and Dufrene, Y. F. *Integrative Biology*, 1, 296–300, 2009. With permission of The Royal Society of Chemistry.)

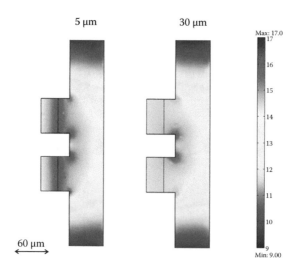

FIGURE 3.16 The logarithm of the gradient of the square of the magnitude of the electric field inside the channel 5 μm and 30 μm above the electrode plane. The logarithm is plotted instead of the gradient in order to visualize the details better. In this example, the total channel height is 60 μm.

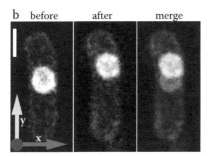

FIGURE 4.11 Figure demonstrating the displacement of the cell nucleus in living *S. pombe* cells. Imaging is enhanced by fluorescence microscopy. The scale bar in the left corner is 3 µm long. Lipid granuli were trapped and moved in a raster-scanning manner to exert a force on the cell nucleus. (Reproduced from Maghelli, N. and Tolic-Norrelykke, I. M. *Journal of Biophotonics*, Vol. 1, pp 299–309, 2008. Copyright Wiley-VCH Verlag GmbH & Co. KGaA. With permission.)

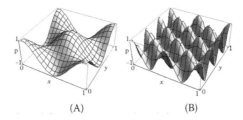

(A)   (B)

FIGURE 5.2 Pressure distribution in a cuboid with rigid boundaries and kz = 0. (A): p = 2, q = 2, r = 0; (B): p = 4, q = 6, r = 0.

FIGURE 5.7 (A) Pressure field plotted for a cross section of an ideal device for the trapping of particles in planes perpendicular to the top and bottom wall. In (B) and (C), the $<p^2>$ and $<v^2>$ terms of the Gorkov force potential—shown in (D)—are represented. The force field is shown as a white arrow in (D) and the trapping planes as a dashed line. Particles are trapped in vertical planes at two locations per wavelength across the channel. For the typical particle fluid combination used in the experiments reported in the literature (e.g., polystyrene beads in water), these locations correspond to the minima of the $<p^2>$ term and the maxima of $<v^2>$ term. If the channel depth h is chosen such that $h<\lambda/2$, the pressure does not vary significantly in the vertical direction and thus no acoustic radiation forces act in that direction.

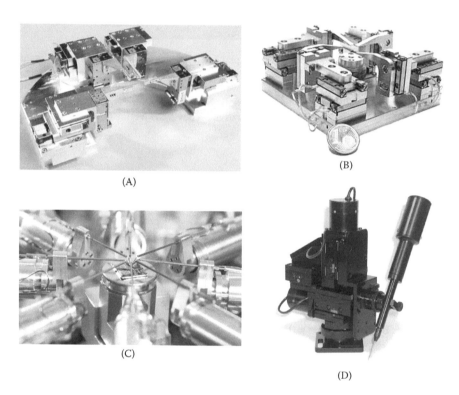

(A)

(B)

(C)

(D)

FIGURE 6.4 Examples of multiprobe systems for micro- and nanomanipulation. Top left: Four micromanipulators from Klocke Nanotechnik (http://www.nanomotor.de/), Top right: A 13-degrees of freedom nanomanipulator from Smaract (http://www.smaract.de/). Bottom left: A multiprobe unit from Zyvex (http://www.zyvex.com/). Bottom right: Piezoelectric micromanipulator from DTI (http://www.dti-nanotech.com/).

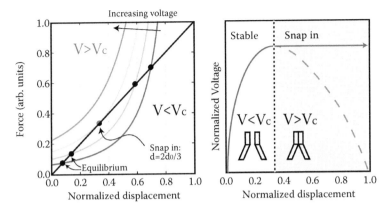

FIGURE 6.6 The mechanical (black curve) and electrostatic (colored) forces are plotted for a voltage-controlled electrostatic actuator. For low voltages, the curves cross at two points, where the lower is the mechanical equilibrium. As the voltage increases, the stable and unstable solutions merge into one at the "snap-in" voltage (blue curve), beyond which there are no stable solutions; that Is, the structure collapses. By plotting the voltage against the normalized displacement, the snap-in occurs at $dV/dg = 0$, or at 2/3 of the initial gap size g_0.

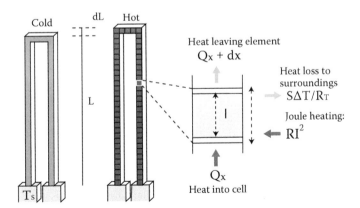

FIGURE 6.9 A U-shaped beam of length L, connected to two room temperature contacts of temperature T_s. The heat flow in and out of the infinitesimal element dx, Q_x and Q_{x+dx}, as well as the heat generated by current heating, RI^2 and the heat exchange with the surroundings, $S\Delta T / R_T$ (see text).

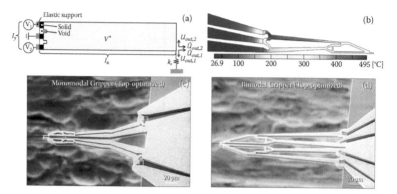

FIGURE 6.12 Topology optimization. The upper panel shows the design domain and the constraints: the overall size, the available amount of material, and the position of electrodes. The goal is to maximize simultaneously the spring constant and the actuation range $(u_{out,1}, u_{out,2})$ of the lower-right corner of the structure. The upper-right corner shows finite element calculations of the temperature distribution using COMSOL. The lower SEM images show a monomodal gripper (can close) and a bimodal gripper (can close as well as open). (Courtesy of Özlem Sardan.)

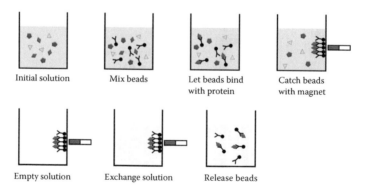

FIGURE 7.1 Schematic of the test tube magnetic manipulation of beads.

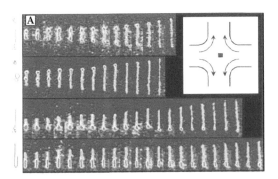

FIGURE 8.3 Stretching of DNA by an elongational flow, visualized by using fluorescently labeled DNA molecules. Each row of images show the unfolding of different molecular configurations in time steps of 0.13 s. The four examples form the classification of configurations, depicted in from of every row and denoted from top to bottom: dumbbell, kinked, half-dumbbell, and folded. (Inset) The channel layout used to create the elongational flow, with the observation region in the middle. (Perkins, T. T., Smith, D. E., and Chu, S. *Science,* 276, 2016–2021, 1997. With permission.)

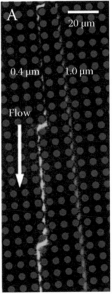

Zagzag Displacement

FIGURE 8.8 Image of a deterministic lateral displacement device sorting fluorescent microspheres, and showing trajectories of the two transport modes. (Huang, L. R., Cox, E. C., Austin, R. H., and Sturm, J. C. *Science,* 304, 987–990, 2004. With permission.)

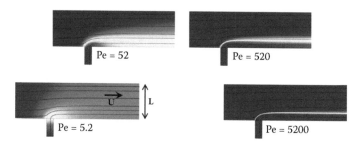

FIGURE 8.9 Numerical example showing the influence of varying Péclet numbers. A side fluid inlet containing a diffusive substance is added to a main fluid flow, and the transition from diffusive transport to advective transport is clearly illustrated, as the Péclet number for the different examples increase in steps of one order of magnitudes. [Numerical simulations done in COMSOL by the author.]

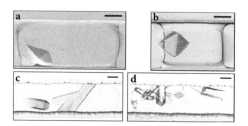

FIGURE 8.18 Polarized light micrographs of protein crystals obtained inside droplets on a microfluidic chip. Scale 50 µm. Reprinted from Zheng, B., Roach, L. S., and Ismagilov, R. F. *Journal of the American Chemical Society,* 125, 11170–11171, 2003. Copyright 2003, American Chemical Society. With permission.)

5 Ultrasonic Manipulation Techniques

Stefano Oberti and Jürg Dual
Institute of Mechanical Systems, Department
of Mechanical and Process Engineering,
ETH Zurich, Switzerland

CONTENTS

5.1   INTRODUCTION

According to historians of science, Robert Hooke of Oxford University is reported to be the first person to use sound to manipulate particles: By using a violin bow and a plate covered with flour, he was able to show the modal patterns of the vibration of the plate already in 1680. Chladni systematically repeated these experiments while replacing flour with sand and published them in his book (Chladni, 1787) titled *Discoveries in the Theory of Sound*. These figures came to be known as the Chladni figures. In 1866, Kundt described a method to determine the speed of sound in fluids and solids, today known as Kundt's tube (Kundt, 1866). Common to all this early work was the goal of understanding sound and vibrations, by looking at how small particles are displaced to nodes of vibration.

Starting with the Euler equations, King in 1934 (King, 1934) computed the acoustic radiation pressure on a rigid sphere suspended in an infinite sound field in an inviscid fluid, that is, the fluid flow is assumed to have no viscosity. The acoustic radiation pressure is a time-averaged pressure, resulting from the scattering of the incident field at the particle. Yosioka (1955) extended the theory to compressible solid particles for the one-dimensional case. The most widely used theory nowadays is based on Gorkov's potential (Gorkov, 1961), where starting with an arbitrary acoustic field distribution the time-averaged force on a compressible spherical particle can be calculated in a straightforward manner in three dimensions. This force is generally called the primary acoustic radiation force and is heavily used in the manipulation of particles.

More complicated situations have been dealt with. Doinikov (1994) and Danilov and Mironov (2000) included the effects of viscosity on the force of a compressible spherical particle. However, the analytical treatment becomes quite complicated. The effects of viscosity are usually neglected, as they are small for particles above some microns.

Because of the complexity of the general situation, numerical methods are gradually being employed despite the enormous computational efforts needed (Wang and Dual, 2009). These methods will allow the computation of the interaction forces, for example, for nonspherical particles, for particles near cavity walls, or for viscoelastic fluids.

If several particles are considered, these particles will interact due to the respective scattered acoustic fields. This will give rise to secondary radiation forces, also called Bjerknes forces (Bjerknes, 1909). These forces can be attractive or repulsive depending on material properties and their orientation with respect to the sound field. They can give rise to the formation of clumps.

Because in standing waves the forces are orders of magnitude larger as compared to traveling waves and because traveling waves are reflected and give rise to standing waves anyway, in most applications, these are used for manipulation tasks and, therefore, often ultrasonic standing wave (USW) manipulation is used.

The frequencies typically used are in the megahertz range, resulting in water of having wavelength of millimeter size and below. The manipulated particles are in the micron range and must have acoustic properties different from the surrounding fluid in order to be manipulated. In early work, biological cells have been manipulated in macroscopic devices for separation science and in bioreactors (Baker, 1972; Coakley et al., 1989; Gherardini et al., 2001; Radel et al., 2000a). Due to the noncontact nature of the ultrasound manipulation, its advantages as compared to other methods soon became clear. Also, it turned out that the viability of the cells is not a problem in most cases, as long as cavitation is avoided (see also Section 5.6)

In Section 5.2, the basic principles will be explained in more detail. With the new manufacturing possibilities in the semiconductor industries giving rise to the field of microelectromechanical systems (MEMS) and micro total analysis systems (μTAS), many new small-scale devices have appeared in the last 20 years. This will be the focus of the device and application sections (Sections 5.3–5.5). The viability of cells will be the topic of Section 5.6, and conclusions are drawn in Section 5.7.

5.2 BASICS OF ULTRASONIC MANIPULATION

5.2.1 ACOUSTICS

For the theoretical analysis of a manipulation device, a two-step procedure is used: First, the acoustic field in the cavity is computed, and second, the force on a particle is computed based on Gorkov's potential.

In the context of linear acoustics and neglecting the viscosity of the fluid, the excess pressure or acoustical pressure p satisfies a three-dimensional wave equation, which is in a Cartesian reference frame with coordinates x_i given by (Kinsler et al., 1982)

$$p_{,ii} = \frac{1}{c^2} \frac{\partial^2 p}{\partial t^2} \tag{5.1}$$

$$c^2 = \frac{B}{\rho_0} \tag{5.2}$$

where $_{,i}$ denotes differentiation with respect to the coordinate x_i, c is the wave speed, B the adiabatic bulk modulus, ρ_0 the equilibrium density, and the summation convention is used for repeated indices.

This is a linear partial differential equation: The classical wave equation in three dimensions, for which many solutions are known in the literature. It allows for dispersion-free propagation of waves with the wave speed c.

Typical properties relevant in this context are given in Table 5.1. The characteristic impedance $\rho_0 c$ of the material is the product of density and wave speed. For plane waves, it is equal to the quotient of pressure and velocity amplitude.

The fluid particle velocity u_i can be obtained from the linear momentum equation

$$\rho_0 \frac{\partial u_i}{\partial t} + p_{,i} = 0 \tag{5.3}$$

It can also be expressed with the velocity potential φ

$$u_i = \varphi_{,i} \tag{5.4}$$

TABLE 5.1

Examples of Acoustic Properties of Fluids

	ρ_0 [kg/m3]	B [N/m²]	c [m/s]	$\rho_0 c$ [Pas/m]
Air at 20°C	1.21	$1.42 \ 10^5$	343	415
Mercury Hg	13600.	$2.53 \ 10^{10}$	1450.	1.97×10^7
Water at 20°C	998.	$2.19 \ 10^9$	1481.	1.48×10^6

Source: Kinsler, L. E., Frey, A. R., Coppens, A. B. and Sanders, J. V. (1982). Fundamentals of Acoustics. New York: John Wiley & Sons.

Note: ρ_0 equilibrium density, B adiabatic bulk modulus, c wave speed, $\rho_0 c$ characteristic acoustic impedance.

The velocity potential also satisfies the wave equation.

When solving Equation 5.1, an initial value problem must be solved that satisfies the boundary conditions. If all these conditions are properly formulated, a unique solution exists.

For particle manipulation, most often a harmonic solution is sought, as the transducers used to excite the acoustic field are driven for times much longer than the typical response time of the system, which is QT, where Q is the quality factor and T the period of the vibration. For a typical system with $Q = 100$ at 1 MHz, the response time is a fraction of a millisecond. We can therefore substitute the time dependence by a harmonic solution of circular frequency ω, that is, all quantities have $e^{i\omega t}$ or $cos(\omega t)$ as a factor, which is understood in the following and often omitted.

The complex description is used whenever damping is involved. In these cases, it is understood that only the real part of the solution has a physical meaning. From Equation 5.1, we obtain for the harmonic case

$$p_{,ii} = -\frac{\omega^2}{c^2}p = -k^2 p \tag{5.5}$$

Here, k is the wave number in the fluid, and the following quantities are defined:

$$f = \omega/2\pi \quad \text{frequency} \quad (1/s, \text{Hz})$$

$$\lambda = c/f \quad \text{wavelength} \quad (m)$$

$$k = \omega/c = 2\pi/\lambda \quad \text{wave number} \quad (\text{radians/m})$$

$$T = 1/f \quad \text{period} \quad (s)$$

For harmonic solutions, the initial conditions are not relevant.

Typical boundary conditions are:

$$\text{a.} \quad \textit{Fixed boundary} \quad u_i n_i = 0 \tag{5.6}$$

This means that the particle velocity normal to the boundary surface (characterized by its unit normal n) must disappear. Note that because the fluid is assumed to be inviscid, no condition is imposed on the tangential velocity.

When formulated in terms of the pressure p, Equation 5.3 together with Equation 5.6 yields for a fixed boundary

$$p_{,i} n_i = 0 \tag{5.7}$$

In reality, all fluids have a certain viscosity, which forces the tangential velocity to be zero at the boundary. This happens within a narrow region called the Stokes' layer.

The thickness of the Stokes' layer is given by the formula:

$$\delta = \sqrt{\frac{\eta}{\rho_0 \omega}} \tag{5.8}$$

where η is the dynamic viscosity. For water at 1 MHz, the thickness δ is about 1 micron, and viscosity might become relevant for very small particles or cavities. The effect of viscosity is usually neglected in acoustics.

$$\text{b.} \quad \textit{Free boundary} \quad p = 0 \tag{5.9}$$

when surface tension is neglected.

$$\text{c.} \quad \textit{Fluid/solid boundary:}$$

When an acoustic fluid is in contact with a solid, none of the above boundary conditions is valid in a strict sense. More precisely, we have

$$u_{iS} n_i = u_{iF} n_i$$
$$\sigma_{n_i} = \sigma_{ij} n_i n_j = -p \tag{5.10}$$

The normal displacements at the interface must be equal, and the normal stress in the solid must be equal to the negative pressure in the fluid. Here, the indices F and S refer to the fluid and solid, respectively, and σ_{ij} is the stress tensor in the solid.

For general geometries, no analytical solutions exist. For simple geometries, we can find solutions by assuming suitable functional dependencies that allow the boundary conditions to be satisfied. This allows us to obtain a feeling for the physics.

A one-dimensional standing wave is described by (Figure 5.1)

$$p = p_0 \sin(k_x x) \cos(\omega t) \tag{5.11}$$

More generally, the one-dimensional solution will look like

$$p = (A\cos(k_x x) + B \sin(k_x x)) \cos(\omega t),$$

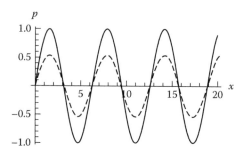

FIGURE 5.1   One-dimensional standing wave according to Equation 5.11 with $k_x = 1$, $\omega = 1$ at two times t: $t_1 = 0$ (solid) and $t_2 = 1$ (dashed).

where A and B need to be determined from the boundary conditions. Standing waves will always occur if waves are reflected from boundaries.

For the homogeneous solution for a three-dimensional cuboid resonator of dimensions L_x, L_y, L_z with rigid boundaries, we might assume a pressure distribution:

$$p = A \cos(k_x x) \cos(k_y y) \cos(k_z z) \cos(\omega t)$$

Using Equation 5.5, we obtain

$$k_x^2 + k_y^2 + k_z^2 = k^2$$

If we take

$$k_x = p\pi / L_x, \quad p = 0,1,2,...$$

$$k_y = q\pi / L_y, \quad q = 0,1,2,...$$

$$k_z = r\pi / L_z, \quad r = 0,1,2,...$$

all the boundary conditions are satisfied, so we have found an infinite number of solutions. The corresponding frequencies are computed from

$$f_{pqr} = \frac{c}{2}\sqrt{(p/L_x)^2 + (q/L_y)^2 + (r/L_z)^2}$$

by using the definition of k.

For micromanipulation, often the z dimension of the resonators is small, such that $r = 0$ and $k_z = 0$.

Figure 5.2 shows two pressure distributions for two values of p, q, and r.

The pressure distributions can have periodic properties, which then result in a periodic arrangement of particles. This periodic arrangement can be in the form of parallel lines or regularly arranged clumps. Note that because of the linearity of the system, any superposition of the aforementioned solutions is also a solution.

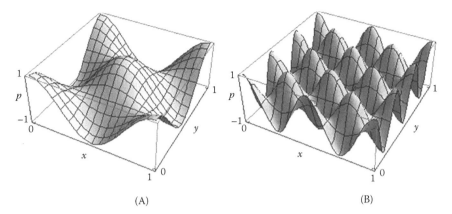

(A)   (B)

FIGURE 5.2 **(See color insert)** Pressure distribution in a cuboid with rigid boundaries and $k_z = 0$. (A): $p = 2$, $q = 2$, $r = 0$; (B): $p = 4$, $q = 6$, $r = 0$.

Because in reality every boundary has a nonzero compliance due to the elasticity of the materials, the foregoing ideal solutions have to be modified depending on the structure surrounding the cavity.

The cavity is enclosed by a compliant structure and transducers are used to generate the field. For modeling, the acoustic equations must be combined with the equations describing the vibrations of the solid and the transducer. Because of the geometrical complexity, usually a finite element solution is computed. The transducers are often made of piezoelectric material (Ristic, 1983) and can be included in the model. They can be used in a resonant mode (narrow frequency range) or nonresonant mode.

5.2.2 FORCES ON PARTICLES

The force acting on a spherical particle can be calculated from Gorkov's potential (Gorkov, 1961)

$$\langle U \rangle = 2\pi r_p^3 \rho_0 \left(\frac{1}{3} \frac{\langle p^2 \rangle}{\rho_0^2 c^2} f_1 - \frac{1}{2} \langle u^2 \rangle f_2 \right) \tag{5.12}$$

where the brackets denote time averaging over one period, r_p is the radius of the particle, and f_i are dimensionless functions containing the properties of fluid and particle:

$$f_1 = 1 - \frac{\rho_0 c^2}{\rho_p c_p^2}$$

$$f_2 = \frac{2(\rho_p - \rho_0)}{2\rho_p + \rho_0} \tag{5.13}$$

where ρ_p and c_p are density and P wave speed of the particle, respectively. In a solid particle, the P wave speed is the wave speed of the dilatational waves.

The primary radiation force is the gradient of the Gorkov potential

$$F_i^{rad} = -\langle U \rangle_{,i} \tag{5.14}$$

For a one-dimensional pressure distribution similar to Equation 5.11, one obtains for the primary radiation force in the x direction:

$$p = p_a \cos(k_x x)\sin(\omega t)$$

$$u = -u_0 \sin(k_x x)\cos(\omega t)$$

$$u_0 = \frac{p_a}{\rho_0 c}$$

$$F_x^{rad} = 4\pi r_p^2 (r_p k_x) E_{ac} \Phi \sin(2k_x x) \tag{5.15}$$

$$\Phi = \frac{\rho_p + \frac{2}{3}(\rho_p - \rho_0)}{2\rho_p + \rho_0} - \frac{1}{3}\frac{\rho_0 c^2}{\rho_p c_p^2}$$

$$E_{ac} = \frac{p_a^2}{4\rho_0 c^2}$$

where $4\pi r_p^2$ is the surface area of the sphere, $r_p k_x$ is the ratio of particle size to wavelength, E_{ac} is the acoustic energy density of the standing wave, and Φ is the acoustophoretic contrast factor, which contains two contributions containing the density ratio and the compressibility ratio, respectively.

If the acoustic contrast factor is positive and no other forces are present, particles will assemble at the nodes of the pressure field (Figure 5.3). Depending on the ratios

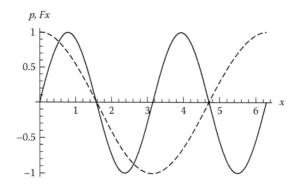

FIGURE 5.3 One-dimensional pressure distribution (dashed) and force (solid) according to Equation 5.15 for $\Phi > 0$: The force pushes the particles toward the pressure nodes.

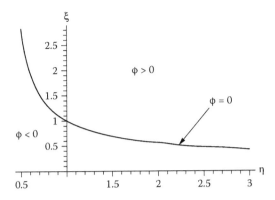

FIGURE 5.4 Acoustophoretic contrast factor as a function of density and wave speed ratios. Most biological cells have $\Phi > 0$.

$$\xi = \frac{c_p}{c}$$

$$\eta = \frac{\rho_p}{\rho_0}$$

(5.16)

the contrast factor Φ can be positive or negative (see Figure 5.4). For a negative contrast factor, the particles move to the pressure antinodes.

Typically, the acoustic pressure amplitude p_a is about 10^5 Pa, such that the energy density will be about 1 to 10 J/m³. Forces for a 10 μm diameter particle at 1 MHz in water are then in the 10 pN range.

5.3 ULTRASONIC MANIPULATION SYSTEMS

In the last couple of decades a rapid increase in the number of publications has been witnessed in the field of ultrasonic particle manipulation, making the identification of the major research directions difficult at first glance. The following sections aim to achieve some clarity. Despite the fact that a systematic classification is hardly possible, we have decided to suggest two possible approaches. On the one hand, in this section, the major designs will be introduced and the devices will be grouped according to the orientation of the trapping planes with respect to the plane of the device (i.e., the top or bottom surface of the fluidic cavity), these being either parallel or perpendicular to it. On the other hand, in Section 5.4, systems will be classified according to their field of application.

It is important to point out here that this chapter is restricted to micromachined systems or microengineered systems whose fluidic cavity dimensions lie below 1 mm. Such a choice is motivated by the fact that ultrasonic manipulation has the potential of being integrated into a *lab-on-a-chip* system (or micro total analysis systems). Furthermore, only manipulation of solid particles (e.g., polymer beads, cells, etc.) is taken into consideration, leaving the discussion of systems handling fluids (e.g., mixers) and bubbles to other reviews. Finally, only those systems are described that use acoustic radiation forces as the primary manipulation mechanism. If acoustic streaming arises in such systems, it is only as a side effect—mostly detrimental—that affects the positioning accuracy or efficiency. Reviews also covering the use of acoustic streaming have been written by Chladni (1787) and Kuznetsova and Coakley (2007). Nevertheless, on some occasions when a similar result was obtained using another technology (e.g., surface acoustic waves, low-frequency vibration, etc.), a short digression will be made.

5.3.1 Design for Particles Trapping in Planes Parallel to the Plane of the Device

A first approach to the manipulation of particles in micromachined systems consisted of the miniaturization of macroscopic acoustic filters. So-called multilayered structures were made of a fluid layer confined between a reflector and a coupling layer (or carrier layer), on whose reverse side a piezoelectric transducer was attached (Figure 5.5). The role of the coupling layer was often that of insulating the transducer from the fluid, rather than giving impedance matching.

When modeled, these systems were often treated as one-dimensional resonators. The thickness of the layers was chosen such that a standing pressure field of half (or an integral multiple of it) or a quarter acoustic wavelength λ could

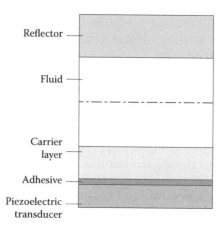

Reflector

Fluid

Carrier layer

Adhesive

Piezoelectric transducer

FIGURE 5.5 Basic design (cross section) of a multilayered resonator consisting of a particle-laden fluid, confined between a coupling or carrier layer and a reflector. On the reverse side of the carrier layer, a piezoelectric transducer is attached. Upon excitation, when the resonant condition is matched, a standing wave pressure field can be set up in the fluid and in the solid structure. A single trapping plane or multiple such planes are then formed, collecting particles either in a single or multiple planes in the fluid, or pushing them toward the top or bottom surface of the fluidic cavity. The dotted line marks the orientation of the resulting trapping plane. The thickness of the various layers is usually chosen to match a half (or a multiple of it) or a quarter acoustic wavelength λ, depending on the desired location of the trapping planes. When modeled, these systems were usually treated as one-dimensional resonators.

be established in the fluid layer and in the solid structure. As a result, particles were expected to be trapped in a plane at half-channel depth (or multiple planes equally spaced [Harris et al., 2003]) or pushed against one of its surfaces parallel to the transducer (Hawkes et al., 2004a; Townsend et al., 2008). An excellent starting point for a detailed description of such systems is the work by Hill et al. (2008).

In the most recent systems, often a polymer gasket was used to laterally confine the liquid and at the same time act as a spacer defining the channel depth (Glynne-Jones et al., 2008; Kuznetsova and Coakley, 2007), thus defining both lateral and vertical dimensions of the fluidic cavity. This way, as the lateral damping was increased, the excitation of unwanted enclosure modes (eigenmodes of the fluid itself) could be avoided. Such modes were identified as the origin of inhomogeneities in the trapping plane. In this regard, Townsend et al. (2006) reported on the detrimental effect that striations formed in the trapping plane had on the separation efficiency of a half-wavelength separator. They caused aggregation of particles with possible following sedimentation. In contrast, the field became more homogeneous across the channel when lateral walls—originally made of Pyrex glass—were reduced in lateral extension or even substituted with PDMS.

The possibility of using different materials (in particular, Macor ceramic instead of silicon) and hence adopt a different fabrication method independent of silicon micromachining facilities was investigated by Townsend et al. (2008). It was demonstrated that the choice of material played a minor role in the determination of the trapping position and of the pressure field amplitude. These parameters were rather determined by the thickness of the coupling layer.

Also, by careful choice of the thickness of the reflector layer (Martin et al., 2005) or of the fluid gap (Hill et al., 2008), the nodal plane location could be arbitrarily set for a determined thickness combination. In particular, particles could be focused close to or at the reflector surface when the reflector itself was in resonance (i.e., when its thickness matched half an acoustic wavelength). Moreover, when the coupling layer and the piezoelectric transducer were in resonance (i.e., their total thickness corresponded to half the acoustic wavelength), a pressure node was formed on the coupling layer surface. In that case it was possible to push particles against this surface of the cavity. By adjusting the layer thicknesses, the node could be displaced within the coupling layer (Harris et al., 2004).

Trapping of particles at half-channel depth was described by Lilliehorn et al. (2005a) as well. In contrast to the systems previously described, in this device the piezoelectric element was integrated in the bottom cavity wall, thus remaining in direct contact with the fluid. The device itself (Figure 5.6) consisted of an SU8 channel (800 µm wide, 90 µm deep) defined on a glass surface, used as reflector, and pressed against a circuit board, where three square transducers (800 × 800 × 200 µm) were mounted. Together with the epoxy layer into which they were embedded, the transducers formed the bottom surface of the channel. This peculiar arrangement seemed to be the cause of an inhomogeneous pressure distribution in the liquid above the excited transducer (Lilliehorn et al., 2005b) giving rise to lateral forces on the order of magnitude of a few hundred of pN (Evander et al., 2007). These forces were used to retain—in a levitated state—flowing particles. Later, the SU8 microfluidic

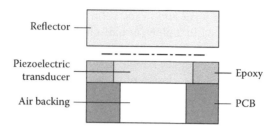

FIGURE 5.6 Schematics showing a cross section of the device design by Lilliehorn et al. (2005b). The piezoelectric transducer (800 × 800 × 200 μm) was mounted on a PCB board and formed the bottom surface of the channel, when embedded into a layer of epoxy. The channel was defined laterally by SU8 structures (not shown in the drawing), later replaced by glass (Evander et al. [2007]) or by silicon (Johansson et al. [2005]). Beside levitation, this particular design originated near-field effects causing strong lateral forces in the plane of half wavelength, trapping the particles in complex patterns. The dashed dotted line shows the location of the trapping plane at half-channel depth.

part was replaced by channels etched directly into the glass reflector (Evander et al., 2007) or by silicon channels (Johansson et al., 2005). In fact, the polymer turned out to be affected by swelling caused by prolonged exposure to ultrasound and by contact with the liquid.

5.3.2 DESIGN FOR PARTICLES TRAPPING IN PLANES PERPENDICULAR TO THE PLANE OF THE DEVICE

Historically developed at the same time as to the ones presented in the previous section, the systems discussed here are characterized by the vertical orientation of the trapping planes with respect to the plane of the device. Acoustic standing fields with pressure gradients in the plane across the channel were set up in these systems. Thus, particles could be trapped in a plane running along the channel centerline or in multiple planes parallel to the sidewalls of the fluidic cavity. When observed from above through a glass plate, the trapped particles appeared as parallel lines or—in most cases—as bands of large numbers of particles. Moreover, the channel depth was often chosen such that for the operational frequencies no significant pressure variation over the channel depth resulted. Thus, no acoustic radiation forces were acting in that direction. As a major advantage with respect to the design presented in Section 5.3.1, it was no longer necessary to match the layer thicknesses to multiples of the acoustic wavelength.

A typical pressure field as well as the related force potential according to the Gorkov theory is plotted in Figures 5.7a and 5.7d. The force field—plotted as white arrows—helps identify the two locations per wavelength at which vertical trapping planes are formed. The $<p^2>$ and the $<v^2>$-components of the Gorkov force potential are shown in Figures 5.7b and 5.7c, respectively. It appears evident that, for an ideal one-dimensional standing pressure field $p = p(x, t)$, such as the one plotted here, it suffices to consider the pressure nodes to know where particles will be trapped. In fact, the component v_x of the velocity vector across the channel dominates in the

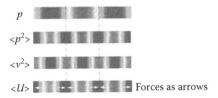

p

$<p^2>$

$<v^2>$

$<U>$ Forces as arrows

FIGURE 5.7 (See color insert.) (A) Pressure field plotted for a cross section of an ideal device for the trapping of particles in planes perpendicular to the top and bottom wall. In (B) and (C), the $<p^2>$ and $<v^2>$ terms of the Gorkov force potential—shown in (D)—are represented. The force field is shown as a white arrow in (D) and the trapping planes as a dashed line. Particles are trapped in vertical planes at two locations per wavelength across the channel. For the typical particle fluid combination used in the experiments reported in the literature (e.g., polystyrene beads in water), these locations correspond to the minima of the $<p^2>$ term and the maxima of $<v^2>$ term. If the channel depth h is chosen such that $h<\lambda/2$, the pressure does not vary significantly in the vertical direction and thus no acoustic radiation forces act in that direction.

force potential calculation (v_z is almost vanishing, and v_y is assumed to be zero) and the velocity is $\pi/2$ phase-shifted with respect to the pressure (i.e., the pressure nodes coincide with the velocity maxima). The role of the $<v^2>$-term in Gorkov's equation is merely that of defining the depth of the potential well, and thus the magnitude of the acoustic radiation force.

In 2003, Dougherty and Pisano (2003) reported on a standing pressure field with a single trapping plane along the centerline of a channel defined by an etched-through silicon spacer (determining the depth of 500 μm, as well as the width 1.5 mm at the top and 0.8 mm at the bottom), sealed on the top by a 175 μm thick glass plate and closed at the bottom by a 700 μm thick Pyrex wafer (Figure 5.8a). The actuation was given by out-of-phase excitation at 1 MHz of two half-channel-wide electrodes patterned on the top surface of a piezoelectric transducer (PZT-5A) bonded on the top glass plate.

The year after, Nilsson et al. (2004) described a simplified excitation for a 750 μm wide, 250 μm deep, and 13 mm long silicon channel sealed by a glass lid (Figure 5.8b). The actuation was produced by a piezoelectric transducer (PZ26) largely exceeding the channel size, glued on the silicon chip reverse side using epoxy and tuned at the frequencies for which standing pressure waves across the channel were set up in the liquid. Up to four parallel bands of particles could be formed under flowing conditions (0.3 ml min^{-1}).

In 2005, Neild et al. (2006b) introduced a third possibility for the creation of trapping planes perpendicular to the fluidic cavity bottom surface. This method was compared later in Neild et al. 2007a to full-plate actuation and the out-of-phase actuation, highlighting its advantages. The so-called strip electrode actuation induced excitation of a wave traveling in a preferential direction in the solid structure, which resulted in a standing pressure field with nodal planes parallel to the excited strip electrode when coupled to the fluid. The system consisted of a 5 mm × 0.2 mm channel, extending 25 mm in length, dry-etched into a 500 μm thick silicon wafer and sealed with a 1 mm thick glass plate glued on its top. On the reverse side of the silicon

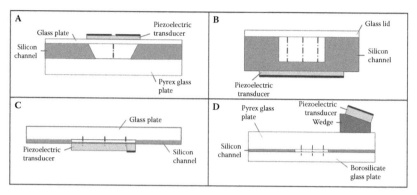

FIGURE 5.8 Schematics of different actuation methods leading to trapping of particles in planes perpendicular to the channel bottom surface (drawings are not to scale). The active electrode is marked by a thick black line (the other electrodes being grounded). The dashed line indicates the orientation of the excited trapping planes. The adhesive used to attach the transducer is not shown. (A) Out-of-phase actuation of a piezoelectric transducer attached to the bottom of the device, for the creation of a single trapping plane, described by Dougherty and Pisano (2003). (B) Tuning of the excitation frequency of a transducer glued beneath a channel of rectangular section, to match the resonant condition in the lateral direction, reported by Nilsson et al. (2004). (C) Actuation of a strip electrode defined on the lower surface of a piezoelectric transducer, in order to force the emission of a traveling wave in the structure, coupling to the fluid in a preferential direction and exciting both symmetric and asymmetric modes, introduced by Neild et al. (2007a). (D) Use of a wedge, onto which the transducer is mounted, aimed at improving the coupling of the transducer displacement into the horizontal direction, described by Wiklund et al. (2006).

channel, a piezoelectric transducer (PZ26) was attached using two-component glue, whereas silver paint served to create the contact for the electrode in contact with the silicon (Figure 5.8c). In contrast to the excitation in Nilsson et al. 2004, rather than exciting the whole transducer, the voltage was applied only to a narrow strip defined on the lower piezoelectric surface (the one in contact with the air), obtained by removing a 200 μm wide strip of conductive layer along the entire transducer length using a wafer saw. The counter electrode, in contact with the silicon, was grounded over the whole area instead. The resulting actuation was asymmetric with respect to the plane passing through the center of the transducer and the channel, so that a larger number of resonant modes could be excited than what could be obtained with actuation of the full plate, which excited primarily symmetric modes (*symmetric* refers to the sign of the pressure, which in this case is the same at both sides of the channel). Both even and odd numbers of particle lines could thus be obtained. The design of the system results to be less complex and facilitates the integration of acoustic manipulation into preexisting microfluidic platforms, where the space might be limited. Moreover, the reduction of the active transducer area corresponded to a minimization of the heat generation during operation. In fact, it turned out that for the typical operating conditions the use of external cooling was not necessary. Finally, with the use of this electrode configuration, there was no need for multiple transducers for the setting up of more complex fields.

A fourth possibility for creating trapping planes along the channel was introduced by Wiklund et al. (2006) (Figure 5.8d). In order to create trapping planes running along a 750 µm wide and 10 mm long channel defined by borosilicate glass plate (150 µm thick), a 40 µm thick silicon spacer and a Pyrex glass plate (785 µm thick), a piezoelectric element (PZ26) was attached onto a refractive PMMA wedge with an angle of incidence of 17° with respect to the plane of the device. This arrangement was aimed at achieving a high coupling of the transducer displacement into the horizontal direction. Moreover, such a design permitted high-NA microscopy, as no transducer hindered the sight across the fluid. In a later work, Manneberg et al. (2008a) investigated the optimum angle by means of impedance spectroscopy and optical analysis of the pattern formed they pointed out that a 30° angle produced the best efficiency, measured in terms of particle concentration in the vicinity of the trapping site.

The last design mentioned here was suggested by Kapishnikov et al. (2006) in 2006 and consisted of half or quarter-wavelength standing fields acting across a 160 and 100 µm wide, respectively, 150 µm deep channel made of PDMS. Excitation was provided by means of two narrow piezoelectric bars located at the sides of the channel embedded in the PDMS. By adjusting their phase a nodal plane was formed along the centerline or on one of its side walls. More details and an application for this system are given in Section 5.4.1. Here, it is noteworthy that this device represents— together with the channel milled in a brass support anticipated by Glynne-Jones et al. (Glynne-Jones et al., 2008)—one of the few examples of quarter-wavelength fields excited across the channel width.

It is important to realize at this point that the designs by Neild et al. (2007a) and Wiklund et al. (2006) offered the possibility of exciting a wave in the solid structure in a preferential direction. This fact is important as it opened the doors to achieving more complex fields by superimposing solid displacements coupling to the fluid in different directions. By doing so, one-dimensional standing pressure fields could be set up also in symmetrical geometries (e.g., square chamber) as well. Two methods of exploiting this peculiarity are mentioned here; others are described in Section 5.4.

Manneberg et al. (2008a) made use of two wedge transducers in order to set up two one-dimensional pressure fields acting across the channel and over its depth, respectively. By doing so, 10.4 µm diameter particles were at the same time focused (at 1.97 MHz) and levitated (at 6.90 MHz) along the channel centerline while they flowed at 5 µl s^{-1}. This transport free of contact with any of the channel walls was more advantageous in terms of sample consumption, as fewer particles remained stuck in the channel. In addition, it provided benefits in terms of microscopy, as the optical focal plane could be made to coincide with the levitation plane. In a later work, the concentration of particles in a clump in at the center of a cage was reported by employing the same method (see Section 5.3.3.2).

An additional peculiarity of wedge actuation lay in the possibility of placing one or more transducers at different segments of the same channel and tune them at different frequencies. Manneberg et al. exploited this idea in Manneberg et al., 2009a and focused their investigation on the extent of the resonance confinement in a segment as well as the presence of the acoustic field in adjacent segments.

Using micrometer-resolution particle image velocimetry (micro-PIV), they could demonstrate that acoustic fields were in fact localized, but the presence of the field could also be detected beyond the channel. In particular, in a channel symmetrical with respect to its centerline, consisting of three segments of varying width of 643, 600, and 500 µm, respectively, the acoustic field was observed leaking beyond the respective segments for a length corresponding to $\lambda/4$. In contrast, in an asymmetric channel of varying width containing split elements to divide particles tracks, the region extended to a length of $\lambda/2$. Finally, upon simultaneous excitation of multiple segments, a sort of negative interference was observed that reduced the amplitude of the force field in the transition region between the two segments (over a length of $\lambda/2$). The importance of this work consisted in the fact that such an approach opens the possibility of integrating different functions on the same chip.

5.3.3 ARRANGEMENT OF PARTICLES IN TWO- AND THREE-DIMENSIONAL AGGREGATES

Trapping of particles in two- and three-dimensional aggregates (also termed *clumps* here) is based on superposition of one-dimensional standing fields acting in two and three different directions, respectively. Both single aggregates and multiple aggregates arranged in a periodic pattern in two dimensions (arrays) have been reported.

5.3.3.1 Arrangement by the Use of Geometrical Constraints

A technique for the creation of a grid of clumps of particles was described by Oberti et al. in Oberti, 2009. It exploited the principle that in a rectangular fluidic cavity, one-dimensional resonances in the two directions in the plane of the device arise at different frequencies. For some frequencies, given a standing pressure field in one direction, destructive interference in the other direction might occur. This was the case when a frequency was chosen such that resonance was matched in one direction, while the wavelength in the perpendicular direction differed by $\lambda_{res}/4$ from the wavelength necessary for resonance in this second direction (where λ_{res} was the wavelength at resonance in the first direction). For these cases, it could be assumed that the fields were almost uncoupled and, in fact, only lines parallel to the first direction were observed in the experiments. Arrays were achieved in two different ways, both requiring actuation of the full plate only, dispensing with the need for strip electrodes. The first way consisted of exciting the piezoelectric transducer alternately at a rate of 20 Hz at two frequencies (2.21 and 2.75 MHz, respectively). As the force fields associated with each frequency were not created concurrently, the resulting overall force field could be found by a further time average of these individual force fields. Hence, the 16 µm diameter particles were collected at the areas in which both force fields had potential minima, or rather as close to that as possible given volumetric restraints. As expected from an analytical model, these locations had a slightly oval shape and different spacing in both directions. As second possibility, leading to the same result, superposition (i.e., sum) of the two frequencies was suggested. The major limitation of this approach was the heat dissipated by the transducer and transmitted to the liquid, which was higher than that generated by a strip electrode. For prolonged operation, this might cause a change in the liquid properties and in turn a change in the wavelength, which do no longer matches the resonant condition.

5.3.3.2 Arrangement by the Use of Actuation

A more powerful approach consists of the exploitation of solid displacements that couple to the fluid in preferential directions and thus set up fields with pressure gradients in different directions. This idea had previously been exploited for a macroscopic system by Haake and Dual (2004), made of a glass plate excited to vibration by means of shear piezoelectric elements glued along its edges and placed over a fluid layer. By means of the excitation of two perpendicularly orientated transducers, a two-dimensional displacement field in the glass plate could be induced. By coupling the vibration to the adjacent fluid layer—not confined by any solid wall on the sides—particles were concentrated in a two-dimensional array of oval-shaped clumps. Two-dimensional positioning of microorganisms in a square chamber has been demonstrated by Saito et al. (2002) by superimposing two standing waves acting in orthogonal directions excited by four cylindrical transducers positioned on the sides of the square chamber.

Arrays of two-dimensional particle patterns have been obtained by Oberti et al. (2007). As a preliminary result, it was shown that lines of particles could be obtained in both directions in a 5×5 mm, 200 µm deep chamber, upon individual excitation of the strip electrode parallel to them. Then, two possible particle arrangements were introduced. When two orthogonal standing waves of exactly the same frequency were excited in the fluidic cavity, two oval-shaped trapping sites per wavelength were formed. Over a larger area, particles appeared trapped in a zigzag pattern (i.e., each oval was at $90°$ to its direct neighbors in either direction). Moreover, an analytical model of the force potential field revealed the contribution of the $<p^2>$ and $<v^2>$-terms of the Gorkov force potential on the overall pattern shape. It turned out that it was the relatively small $<v^2>$-term (for the case considered, only 3% of the maximal amplitude of the $<p^2>$-term) that kept the particles in adjacent clumps apart. Figure 5.9a reproduces the modeled force potential field for one wavelength in each direction. This type of pattern was previously obtained with HeLa cells by Haake (Haake, 2004). The result revealed the limitations of this approach for applications dealing with biological material where cross-contamination between clumps had to be avoided. In fact, contact between two clumps could not be excluded a priori when high-concentration samples were used. This was, however, not the case when a frequency difference was introduced. The shift had to be small enough that the same mode was excited in both directions. On the other hand, it had to be large enough that fluctuation of the force field between the previously described field and one with the clumps oriented in the opposite direction (i.e., "x" instead of "$\Diamond$") did not occur. As the frequency difference was made smaller, the fluctuation rate increased up to the point where the particles were no longer able to follow the change in the field and—in a sort of further time-averaging effect—remained trapped in the location common to all force fields, that is, in a round clump at the center of this locations (Figure 5.9b). In contrast to the previous case, here it was the relatively large $<p^2>$-term that determined the shape of the clump. The $<v^2>$-term—having the same

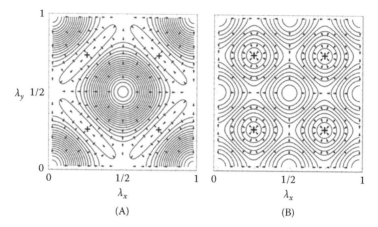

FIGURE 5.9 Force potential field resulting from the superposition of two orthogonal one-dimensional standing pressure waves, plotted for one wavelength in both directions. (A) Oval-shaped trapping sites (centre marked by a "+", two per wavelength in both directions) are formed when the two one-dimensional standing waves have the same frequency. (B) Circular trapping areas (center marked by a "+") are formed instead, when a shift is added between the two frequencies (the same vibration mode still being excited). The analytical model presented by Oberti et al. (2007) proved that the circular aggregates bypassed the limitation set to oval-shaped clumps, which might merge together causing cross-contamination in the case of biological samples.

shape and the location of the maxima coinciding to the minima of the $<p^2>$-term—contributed only to further increase the depth of the potential wells. For the reported experiment, 25 Hz frequency difference were applied between the two signals (at around 2 MHz).

Using the same principle, Neild et al. (2007b) showed the two-dimensional arrangement of MCF10A cells. The time required for positioning was higher, as a consequence of the different material properties of cells.

Manneberg et al. (2008b) reported on the arrangement of particles and cells in lines, in monolayers and in three-dimensional clumps in a $300 \times 300 \times 110$ μm cage defined in a glass–silicon–glass structure similar to the one described previously, infused through an inlet of 110×110 μm cross section. One wedge transducer operated at 2.57 MHz laterally confined the particles in a plane passing through the cage centre, in the inlet direction, whereas they were levitated by exciting a second transducer at 6.81 MHz. In such a configuration, the system was used for long-term enrichment of 5 μm diameter beads into aggregates of up to 96 beads and for arraying of HEK cells in a chain of four. Furthermore, the system offered the possibility of transforming three-dimensional structures into two-dimensional monolayers, by decreasing the voltage of the transducer at 2.57 MHz, thus reducing the amplitude of the acoustic field. Further applications of this system are reported in Section 5.4.6.

Positioning within a micromachined system has been anticipated by Lilliehorn et al. (2005a), who suggested the possibility of creating two-dimensional arrays of coated bioparticles or cells for cell-based assays. Nevertheless, this seems to have never been fully implemented. In the literature, only a series of three transducers has been reported (Evander et al., 2007; Lilliehorn et al., 2005a).

5.3.4 PREDICTION OF NONIDEAL SYSTEM BEHAVIOR

The experimentally observed behavior of the systems described earlier often could not be explained by simplified models such as an ideal one-dimensional resonator (for the systems presented in Section 5.3.2) or such as a rigid resonating fluid volume (for the systems in Section 5.3.3), where the driving frequencies were determined by simply equating $n.\lambda/2 = d$, λ being the acoustic wavelength, n an integer, and d the channel dimension (e.g., width or depth). The real behaviour could be understood only by considering the whole system, in particular by not restricting the analysis to the fluid volume only but by taking the fluid–structure interaction into due account. With the knowledge gained in the last years in this area and the progress in numerical methods (especially finite element simulations), an increasing number of research groups have focused their attention on these aspects. It is nowadays widely recognized that irregularities in the force fields and particle patterns are often caused by a complex behavior of the whole structure and the actuation.

As already described in Section 5.3.1, through eigenmode simulation of a channel cross section, Townsend et al. (2006) could attribute the cause of inhomogeneities in the trapping plane of a multilayered resonator to the lateral boundary conditions, namely, the extension as well as the material properties of the side walls of the channel.

Neild et al. (2006b) demonstrated that for systems with trapping planes perpendicular to the top and bottom surface, it was not sufficient to consider the fluid resonating between rigid walls for a correct prediction of the resonant frequencies. They showed that by means of a two-dimensional frequency response analysis including the fluid–structure interaction, the experimentally used frequencies were matched in a range between 1.0 and 1.7 MHz, with the highest deviation amounting to 0.02 MHz. One of the major outcomes of that work was that the periodicity λ_x of the pressure field across the channel (x being the direction) is given by $\lambda_x = c_x/f$, c_x being the phase speed in the x-direction for the coupled system (considering both the fluid and the structure).

In a further work on a 1 mm wide, 200 μm deep channel ([Neild et al., 2006a]; see Section 5.4.6 for more details and an application), a similar two-dimensional model revealed that particle positioning in a plane along the centerline was possible at two distinct frequencies (this would have not been recognized if only a resonant fluid was considered). Being the model restricted to a channel cross section, it was, however, not possible to further investigate the reason. An explanation was given by Oberti et al. 2008 when the system response was studied using a three-dimensional model, including pressure release boundaries at both ends of the channels. The simulation revealed that at 727.5 kHz a single trapping site confined longitudinally along the center of the channel was formed (Figure 5.10a). It attracted particles from the front and rear end of the channel toward its center. At 750 kHz, a second frequency was found that in contrast was characterized by two distinct trapping sites along the centerline (Figure 5.10b). The detrimental effect of the forces acting along the channel (the

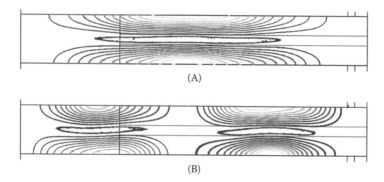

(A)

(B)

FIGURE 5.10 Force potential field resulting from a three-dimensional finite element simulation (frequency response analysis), based on the work of Oberti et al. (2008). At both left and right boundaries $p = 0$ was set (on the right end of the channel, the entrance of the two side channels can be seen). As the channel depth was chosen such that no significant pressure gradient was formed over the channel depth, only slices in the device plane can be considered. (A) At 727.5 kHz, a single trapping site is formed, confined between the two pressure release boundaries at both ends of the channel (on the top and bottom in the figure). (B) In contrast, at 750 kHz, two distinct trapping locations are expected. These differences can only be predicted using a three-dimensional model including fluid-structure and would not be observed in flow-through operation of the system.

so-called lateral forces) will be discussed in Section 5.4.6. It is important to realize that in a flow-through operation, these differences would not have been observed.

Several authors reported irregularities over long, narrow channels. Manneberg et al. (2008a) compared experimental observations and results from two-dimensional finite element simulations of acoustic fields excited at around 2 MHz, the frequency for the generation of a half-wavelength field across the channel. The formation of distinct trapping locations along the axis was attributed to eigenmodes of the whole structure. The same group, in a more recent work (Manneberg et al., 2009b), drew attention to the fact that over a narrow frequency range (6.85 to 6.95 MHz) several frequencies existed for which particles were trapped only over a short length of the channel and that these regions differed for different frequencies. Such an observation led them to develop a method for stabilizing the field, in order to obtain a continuous line over long distances. The frequency was linearly swept at 1 kHz from 6.85 to 6.95 MHz, so that the regions of poor field present at each frequency were averaged out with the better fields present at the same location for other frequencies. Hagsäter et al. (2008) investigated by means of finite element simulations the origin of such discrete trapping sites, concluding that they result from a complex interrelationship between eigenmodes of the fluid only and modes of the coupled structure. Nevertheless, being based on a simplified approach with some assumptions, no precise details could be given.

As a final remark, it must be noted that even for systems where the actuation was in a preferential direction, frequencies existed for which two-dimensional complex patterns, due to eigenmodes not being sufficiently damped, were observed (see for instance [Oberti, 2009]). Manneberg et al. (2008a) exploited this outcome to position clump particles in an array, similar to the one described by Oberti et al. (2007). They observed that a frequency existed at which force fields with components both along and across the channel were set up. Thus, by exciting the device at this single frequency, an array of clumps was formed in the channel. However, the pressure field was characterized by imperfections and missing clumps (where they would be expected in a perfect two-dimensional arrangement) or elongated clumps.

5.4 APPLICATION FIELDS FOR ULTRASONIC PARTICLE MANIPULATION

5.4.1 FILTRATION, CLARIFICATION, SEPARATION

As seen in Section 5.2 the acoustic radiation force acting on a particle depends—among other factors—on the particle size (i.e., volume), its density, and its compressibility. Therefore, according to these two criteria, particles can be separated. Separation of polystyrene particles from the carrier medium where they are suspended (i.e., filtration) in a microengineered stainless steel chamber has been described by Hawkes and Coakley (2001). Particles were focused at the center of the chamber (10 × 0.25 mm cross section, 110 mm long) sealed with a 2.5 mm thick stainless steel reflector. A silicon gasket was placed along the edges of the chamber, between the fluid and a spacer used to define the chamber height. The particle-free medium was removed 10 mm downstream through a slot leaving the main channel at an angle of 9°. With this configuration, a 1000-fold clarification of 5 μm diameter particles was achieved.

Separation in a silicon-glass device operated at half-wavelength mode was reported by Harris et al. (2004) and later by Townsend et al. (2008) (Figure 5.11). By pumping fluid into an inlet defined in the silicon substrate (at a typical flow rate of 200 μL/min) and removing it at different ratios through two separate outlets, particles could be focused at half- (Harris et al., 2004) or quarter-channel depth (Townsend et al., 2008) in a channel etched in Pyrex glass. This way, particle-free fluid could be removed from the most upstream outlet while all the particles were seen leaving the device from the second outlet. In the same work, Townsend et al.

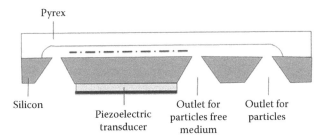

FIGURE 5.11 Schematics of the separator by Harris et al. (2004) and Townsend et al. (2008) (not to scale). Particles were focused either at the center (in the vertical direction) or close to the Pyrex reflector, in order to separate them from the carrier medium. Particles could therefore be removed through a different outlet with respect to the clarified medium. In the figure, only the trapping plane in the middle of the channel is shown, by a dashed dotted line.

presented a modified design consisting in a micromilled device made of Macor. The factors detrimental to performance were studied, these being the position of the trapping plane relative to the reflector, the uniformity of this plane (e.g., formation of striations), and the lateral acoustic radiation force acting against the fluid and causing trapping of particles within the device.

Hawkes et al. (2004a) also investigated the continuous field flow fractionation of aqueous suspensions of yeast cells in an ultrasonic chamber. The cell suspension in a sodium fluorescein solution and clear water were inserted and removed from two different inlets and outlets (at 90° with respect to the main chamber) of a 51 mm long chamber (cross section 0.25 × 10 mm) made of stainless steel, defined by lateral PDMS walls and sealed with a quartz glass reflector, clamped together. As the flow of cells was confined to a small layer close to the steel walls, in the absence of ultrasound, no transfer of particles into the other fluid occurred. By application of ultrasound, cells were forced toward the nodal plane at the center of the chamber and thus transferred to the clear medium. An up to 42-fold increment in the transfer of yeast cells against a small increment of less than 1.2% in the sodium fluorescein was observed upon activation of the ultrasound, as the latter was less affected by the acoustic radiation force due to the reduced size. After a first increment of transfer by raising the voltage, when the voltage was increase above an optimum of 18 V_{pp}, cells were seen to clump, disrupting the flow patterns and thus reducing efficiency.

Gonzalez et al. (2010) described the separation of 6 and 20 µm diameter polystyrene beads under different flow regimes (5.95 µL/min to 28.8 µL/min) for durations up to 15 min and different particle concentrations. An asymmetric design—similar to the previously described one by Hawkes et al. (2004a) but made of polymers—was chosen of two different inlets and outlets that were used, on one hand, to inject the carrier medium and the particle suspension, and, on the other hand, to remove the particles distinctly. As a consequence of the materials used. It consisted the trapping location was not found at the center of the 395 µm wide channel, but rather 117 µm away from the reflector surface. In the other direction, the channel (depth of 250 µm) was kept smaller than a quarter of wavelength in order to avoid the excitation of unwanted modes. Separation above 80% has been reached in all the cases, by operating the device at frequencies around 1 MHz.

Forensic analysis of sexual assault evidence was reported by Norris et al. (2009) as a potential application (acoustic differential extraction) for a system based on the design by Lilliehorn et al. (2005b). Exploiting the size difference between sperm cells and female epithelial cell lysate, that is, by retaining the first while not affecting the latter, separation of the two from a sample of sexual assault evidence could be obtained. The original design was extended with hydrodynamic focusing as reported by Evander et al. (2007) to increase the trapping efficiency and was designed so as to have a channel depth corresponding to $3\lambda/2$, in order to guarantee bigger trapped volumes. During operation, the unretained female lysate was directed toward a first outlet, using a flow ratio created by injecting fluid from a side arm. After completion of the separation (14 min), sperm cells were released and removed from the channel through a second outlet. Further analysis was done off-chip using standard processing methods. The purity achieved in this work corresponded to the values reached with standard methods, bypassing the time-consuming steps of centrifugation and wash steps of standard methods.

Moving to systems with trapping planes perpendicular to the top and bottom walls, the work by Nilsson et al. (2004) is noteworthy. Particle enrichment was proposed as an application, by gathering particles in two bands leaving the channel through the two outer branches of a trifurcated outlet. To this end, an aqueous solution of 5 µm diameter Orgasol particles (Doppler fluid) was used. Particles were focused along the edges of a channel and left the main channel through the two outmost arms of a trifurcated outlet. Two designs were taken into consideration, differing in the angle at which the outlet side arms left the main channel. The study showed that, in general, an increment of the experimental parameters led to poorer efficiency. This was due to the shorter exposure of the particles to the field at higher flow rate and to the formation of clumps at higher voltages, which caused clogging of the channel. From a comparison of the two designs, it was concluded that the chip with 45° outlets achieved higher separation efficiencies than the one with 90° outlets, and this was attributed to the better flow profile.

The design by Nilsson et al. (2004) was used as a starting point in the following work by the same group. In 2005, Petersson et al. (2005a) reported on bovine blood separation into its component erythrocytes and lipids using the same system. Lipid particles are responsible for disorders after cardiac surgery and, therefore, it would be of advantage to separate them from the rest of the blood. The experiment took advantage of the spatial separation due to their different mechanical properties when exposed to ultrasound. While the erythrocytes were focused along the centerline and exited through the central outlet, lipids gathered at the antinodes (i.e., the regions of maximum pressure fluctuation) along the channel walls and left through the two external branches.

In order to face the high throughput volumes required by applications such as cardiac surgery, eight (or more) channels have been operated in parallel (at 0.5 mL/min) by the use of a single transducer driven at around 2 MHz as described by Jönsson et al. (2004). In this configuration, 60 mL/h can be processed, which would however still require a 20-fold increase to meet clinical demands. The separation of lipids reached values between 66% and 94% with an average of 81%.

Without leaving the field of clinic diagnostic applications and biomedical analyses, it must be added that the generation of plasma of low cellular content is often of particular importance. This is done by removing enriched blood cells from whole blood. For the handling of such high-concentration samples, high acoustic radiation forces are required in order to focus the particles in a narrow band. This often represents a problem given the limited forces that can be achieved (this being the case when the volumetric ratio exceeds 5%–10%). Lenshof et al. (2009) solved this problem by prolonging the channel in meander, so that the flowing cells (80 µL/min) were exposed for a longer time to the ultrasonic field. Channels of 56, 108, 166, and 224 mm length were tested. In these systems, enriched blood cells were sequentially removed from a channel through a series of outlets placed on the bottom surface along the centerline, decreasing their concentration in the channel stepwise. In the experiments, the longer channels showed good efficiency also at higher hematocrit concentration (around 40%, corresponding to whole blood). Nevertheless there always were a relatively high number of cells in the plasma fraction. In fact, only with the longest channel a value below the threshold of erythrocytes recommended for plasma transfusion by the Council of Europe (6.0.10^9 cells/liter).

In 2006, Kapishnikov et al. (2006) released results on dilution of particle suspension in a one- or three-stage mechanism as well as size-selective particle sorting of mixtures of two sizes. Three devices were built for these purposes, all made of PDMS. In the first system, three pairs of side arms were placed along a meander-type channel (cross section of 160 × 150 µm). Particles were forced by acoustic radiation to leave the main channel through the side outlets (by placing a trapping plane along the sides of the channel), whereas the diluted solution flowed to the next dilution step through the central outlet. Continuous blood cell separation from plasma was reported as an application. A separation efficiency up to 99.975% was achieved (given as a percentage of the ratio between the initial and final concentration in inlets and outlets, respectively). On the other hand, for size-selective separation, a system with two inlets and two outlets (cross section 100 × 120 µm) was built, which was designed according to the principle that larger particles could be moved faster toward a flow of pure solvent, entering from the other inlet, than small particles, which are less affected by the acoustic field. Thus, large particles exited through the solvent outlet, while the smaller ones left the channel through the other outlet. In order to achieve this, the trapping plane was made to coincide—by phase adjustment—with the wall on the pure solvent side.

The work by Kapishnikov et al. (2006) gives us an opportunity to mention another area of particle separation.

Petersson et al. (2007) reported on separation of a mixture of particles of several sizes. When the drag force in a laminar flow was superimposed to the particle-size-dependent acoustic radiation force, particles were collected at different locations across the channel. Larger particles were displaced faster toward the central plane and were collected in the central outlet, while smaller particles flowed down the channel closer to its sidewalls. All these particles left—in different streams—through two side channels, which were further bifurcated in a total of eleven outlets, each stream ending up in an outlet. Separation efficiencies between 62% and 94% for 2, 5, 8, and 10 µm particles were observed. Furthermore, separation of particles with similar densities was reported in the same work. By setting the density of the carrier medium to a value between the densities of similar particles, erythrocytes, platelets, and leukocytes could be separated.

In 2008, Evander et al. (2008) reported on the extraction of 5 µm diameter polyamide particles performed in an all-glass device, where a 375 µm wide (corresponding to $\lambda/2$ at 2 MHz in water), 125 µm deep, and 30 mm long channel was isotropically etched into a glass wafer. In comparison to a silicon channel of the same design (but different cross section), at higher flow velocities the performance of the glass chips was, however, inferior to that of the silicon one. In regard to particle concentration, no significant difference was observed; instead, for both designs the separation efficiency decreased with increasing particle concentration.

The same design was used by Laurell et al. (2007) for size-sensitive separation of mixtures of particles of two different sizes. This was achieved by switching between three different frequencies. As particles of different size travel at different speeds due to the different acoustic force experienced, after some switching steps between a frequency resulting in one and a frequency resulting in two trapping sites across the channel, the large particles found themselves in a region closer to the central plane whereas the small ones were closer to two external planes. This way, they could flow out of the channel through different outlets.

5.4.2 CARRIER MEDIUM EXCHANGE

Devices based on the design presented by Nilsson et al. (2004) with three inlets and outlets were reported later by Petersson et al. (2005b) for carrier medium exchange applications. By injecting clean medium through the central inlet and contaminated medium through the two side inlets and by exciting a half-wavelength acoustic standing field across the main channel (cross section 125 × 350 μm), particles were pulled from the contaminated fluid flowing along the channel walls into the clean one and confined at the center, after having been exposed to ultrasound at about 2 MHz along 30 mm of the channel length. On the opposite side, particle-laden clean medium was removed from the central outlet, while contaminated particle-free medium left the channel through the two side branches. 95% of the particles could be transferred into the fresh medium. Furthermore, it was noticed that the medium exchange efficiency decreased with increasing voltage, this being likely due to mixing as a result of violent particle movement caused by higher acoustic radiation forces. Finally, by introducing a clean-medium buffer zone along the two external outlets, the efficiency could be further improved. Transfer of erythrocytes from contaminated into fresh blood plasma (blood washing) using this method was described as well.

Carrier medium exchange can be obtained in the opposite way as well. Rather than displacing the particles, the original carrier medium can be removed from the channel while fresh one is injected in the same amount. Preliminary results of a sequential buffer exchange system suitable for low-density suspensions of large particles has been reported in 2008 by Augustsson et al. (2009). In a 6 cm long, 375 μm wide, and 150 μm deep channel, particles or erythrocytes injected at 50 to 120 μL min⁻¹, depending on the experiment, were focused along the channel center-line (operation at λ/2 at 2 MHz). At the same time, particle-free buffer was aspirated through eight side channels (width 150 μm), while new buffer was inserted in the same amounts from the eight channels located on the opposite side. Close to the side channels, the path of particles was disrupted toward the outlet—this being caused by the fluid flow—but it was regained downstream, prior to meeting the next junction 6 mm farther. The efficiency was found to be up to 96.4% in the case of 5 μm diameter polystyrene beads and 98.3% for erythrocytes, respectively.

5.4.3 MODIFICATION OF THE PARTICLES' TRAJECTORY: VALVING AND SWITCHING

An asymmetric design, as for instance the one by Kapishnikov et al. (2006) described earlier, can be used as the valving system. This has been suggested by Laurell et al. (2007) (Figure 5.12). Instead of injecting two fluids, the system was operated with the same fluid. Particles entering from one inlet flowed out of the channel

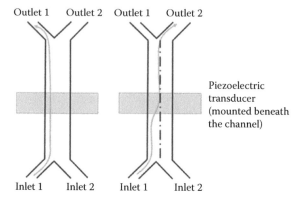

FIGURE 5.12 Schematics of the valving principle described by Laurell et al. (2007). By designing the inlet and outlets asymmetrically with respect to the centerline of the main channel, the trapping plane set up upon activation of the piezoelectric transducer (located at the backside of the channel) moved particles slightly more toward the side of outlet 2. As a consequence, particles that flowed close to the left wall when no sound field was applied (a) were forced to leave through the opposite outlet upon excitation of a half-wavelength field across the main channel. The particle trajectory is shown in gray.

from the outlet on the opposite side when a half-wavelength field across the width was set up, or from the one on the same side when the transducer was not excited. Such switches could be combined in order to achieve in-plane switching networks, working upon activation of different channel segment (with different width) actuated at different frequencies. In addition, different functionalities (such as trapping) could be integrated at different locations of the branched channel. According to the authors such a design could find application in the selection and counting of cells, identified by specific antibodies, and use it as a fluorescent-activated cell-sorting device (FACS).

In this context, it is worthwhile remembering, that the device by Manneberg et al. (2009a) presented in Section 5.3.2 can be used to merge streams of particles in different segments of a channel containing split elements.

5.4.4 BIOASSAYS

In 2004 Hawkes et al. (2004b) investigated the possibility of enhancing the capture of *Bacillus subtilis var niger* spores on an antibody-coated glass surface, overcoming in this way the limitation of diffusion-limited systems. This surface formed the reflector (1 mm in thickness) of a flow-through (flow rate 0.2 ml min⁻¹) resonator made otherwise of a stainless steel layer (carrying the piezoelectric transducer on the air side) and a fluid chamber delimited by a silicon-rubber gasket. The chamber height was 178 and 200 μm for the two built chips, respectively. By operating the system at the reflector resonance, a pressure node was formed close to the reflector wall. Spores, which in absence of the ultrasonic field did not attach to the substrate, upon activation were seen to first form clumps and then form a monolayer on the reflector surface. A 200-fold enhancement of spore capture could be achieved.

More recently Glynne-Jones et al. (2010) reported on a novel idea in microbeads assays based on multi-modal excitation. A mixture of functionalized streptavidin-coated microbeads (polystyrene, 6 µm in diameter) and nonfunctionalized control beads (6 µm in diameter) was first pushed against a functionalized glass surface (PEG-biotin), by applying a quarter-wavelength mode with a nodal plane located into the reflector (rather than at its surface in contact with the fluid) in order to overcome repulsive electrostatic forces. Then, after 90 s, the control beads were released by switching to another quarter-wavelength mode with a trapping location on the opposite side on the chamber, namely, at the interface between the fluid and the carrier layer. Finally, the control beads were removed by shortly flushing the chamber. As a comparison, without application of ultrasound after 270 s only a small proportion of the beads were bound.

Evander et al. (2007) reported on perfusion applications. In such a 600 µm wide 61 µm deep channel, based on the principle of Lilliehorn et al. (2005a) and operated at 1.24 MHz, yeast cells have been successfully trapped and perfused (1 µL/min with cell medium to promote their growth) for 6 h. In order to increase the trapping efficiency, the stream of particles was hydrodynamically focused at the center of the channel as it entered it. Another perfusion application is described later in Section 5.4.6.

5.4.5 PARTICLE ENCAPSULATION IN A MATRIX

Gherardini et al. (2005) reported on a cell immobilization technique for encapsulating latex beads into a gel matrix (technical agar gel). In a 250 µm wide microengineered channel, particles were collected in the trapping plane formed at 3 MHz and maintained there by keeping the ultrasound on. At the same time, the temperature was decreased by placing the device in an ice bath, leading to a solidification of the gel. After 15 min of ultrasound application, the channel was opened, and the solidified gel removed. The experiment proved that particles maintained their position during gel solidification. With this approach, an encapsulation technique for biological samples was presented, which can be used for the preparation of an environment for the development of cell colonies and for cell viability studies.

Seeding of cells into scaffolds was reported by Li et al. (2009) using surface acoustic waves (SAW) as well. A scaffold (4 × 4 × 2 mm) was placed on a substrate close to an interdigitated transducer (IDT) and a droplet of osteoblast-like cell suspension was then deposited between them. The entire scaffold was seeded within 10 s (in contrast to standard diffusion-based methods, requiring from 30 min up to days), with a comparable or better cell distribution than the latter. The study was focused on the investigation of viability, proliferation, and differentiation of cells. It was ascertained that exposure to a 20 MHz SAW did not significantly affect the cell viability, but when exposed to 10 MHz SAW, only half of them survived. This was attributed to the larger displacement generated at lower frequencies. On the other hand, proliferation and differentiation did not show differences form those measured for untreated cells.

5.4.6 SELECTIVE OR CONTROLLED MANIPULATION OF SINGLE PARTICLES

Acoustic manipulation is by its nature a multiple particle manipulation technique, as particles all react simultaneously upon activation of an acoustic field. Furthermore, the force potential gradients are usually not steep, so that high accuracy in the position cannot be achieved. All the systems and applications discussed earlier exploit this principle. Nevertheless, some significant efforts have been made in the last couple of years toward single-particle manipulation. Two main directions have been investigated. On the one hand, perfusion chambers have been reported, in which single particles were loaded and trapped. As acoustic manipulation is not able to distinguish between particles, this technique requires that single particles be singularly loaded into the channel. On the other hand, acoustic radiation forces have been used in a preparation step prior to mechanical manipulation. Particles were positioned at known sites (so that localization was no longer necessary), in such a way that they were individually accessible to the tool used for further manipulation.

Using wedge transducers, Manneberg et al. (2009b) managed to move particles along a channel (cross section 110×110 µm) at a flow rate of 0.2 mm s^{-1} and finally insert them into a $300 \times 300 \times 110$ µm cage without a net fluid movement. Only the trapping sites were displaced along the channel. Toward this end, the excitation frequency was linearly swept from 2.60 to 6.64 MHz at a rate of 0.2 Hz (experiments were conducted at 0.5 and 0.7 Hz as well, increasing the speed of the particles or particle clumps along the channel). The principle is based on the fact that the particle at the end of the sweep lies closer to the next node of the starting frequency than the one at the beginning of the sweep. Moreover, in order to stabilize the acoustic field in the injection channel, the frequency was linearly swept from 6.85 to 6.95 MHz at the rate of 1 kHz, as mentioned before in Section 5.3.4. Transport of clumps was demonstrated as well. However, in the case of high-concentration samples, some particles were seen leaking from one clump and flowing into an adjacent one. With this principle, the system for intercellular interaction reported by Manneberg et al. (2008b) could be extended.

Using three transducers and the same actuation principle, Svennebring et al. (2009) reported on the selective injection and retention of cells and other bioparticles in a horizontal monolayer in a confocally shaped perfusion chamber (Figure 5.13). One wedge transducer was located along the 328 µm wide inlet channel, one in proximity of the chamber and one extended over both the channel and the chamber. The device integrated a set of manipulation functions. The chamber was continuously excited at 7.07 MHz. In this clever design, this frequency, besides creating a retention zone at the center of the resonator, also corresponded to half wavelength in the chamber's vertical direction, so that the particles were eventually trapped in a levitated state at half-chamber depth. In order to inject particles into the chamber, the inlet channel was excited at 2.1 MHz, which focused them in a line coinciding with the channel centerline. When the frequency was switched to 4.60 MHz instead, two nodal planes were created and particles in the channel could flow through the chamber without passing through its center (bypass function). The device was operated at a constant flow rate of 5 µL/min. The authors have identified several areas

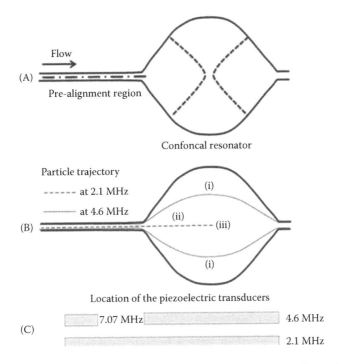

FIGURE 5.13 Schematics (top view) of the confocal resonator by Svennebring et al. (2009). (A) Acoustic fields excited within the fluidic cavity upon excitation: in the inlet particles were prealigned along the central axis (both in the lateral and vertical directions, dotted line) by activating the 7.07 MHz transducer, whereas a confocal field (dotted line) was set up in the cavity (excited by the 4.6 MHz transducer). (B) Manipulation functions: (i) by pass function, where the particles were free to leave the confocal resonator without being trapped, (ii) injection in the retention region (focus of the resonator), and (iii) retention. (C) Location and operational frequencies of the three transducers.

where this device could find application: First, in the enhancement of bead-based immunoassays. Second, in the real-time investigation of the response of assemblies of 10 to 100 cells to varying parameters (e.g., concentration of chemicals), where the number of cells can be controlled using the bypass function. Third, the device is suitable for the study of parameters with slow dynamics (i.e., hours to days), such as cell activation, differentiation, and proliferation or such as the monitoring of immunological synapses for studying immune cell interactions. Finally, it has to be said, as reported by the authors, that manipulation of single particles with this system is in principle possible, but the low flow rates would limit the efficiency to a few particles per minute.

Schwarz and Dual (2009) recently expanded the possibilities of controlled particle manipulation by adding rotational capability to the trapping, thus opening new opportunities for construction and positioning of microcomponents. The technique relies on the fact that nonspherical particles are subject to a torque that can be steered with a time-varying pressure field. Thus, by amplitude modulation of two orthogonal standing waves, the orientation of the force potential minima could be controlled, varying from a "◊"-shaped potential to an "x"-shaped one passing through a force potential characterized by two parallel lines. In a 3 × 3 × 0.2 mm chamber etched in silicon and sealed on the top by a glass plate, 180° rotation of a 200 µm long fiber was achieved. Toward this end, first the amplitude of one standing wave was changed by varying the voltage applied to the corresponding strip electrode of the piezoelectric transducer (while keeping the other at a constant value), and vice versa thereafter. The orientation of the fiber could be stopped or even reversed.

Beside these systems—as mentioned earlier—another direction was followed. It consisted of using acoustic fields to preposition particles in such a way that they could be individually accessed, thus acting as an important preparatory step toward automated mechanical manipulation. Such systems must feature an interface through which the tool can be inserted in the fluid. In 2006, Neild et al. (2006a) described the semiautomated sequential removal of 74 µm diameter copolymer particles from a channel (1 × 0.2 mm in cross section) using a capacitively actuated microgripper. In order to achieve accurate positioning, particles were first roughly collected along the central axis of the channel by exciting the system at 780 kHz, in order to ensure that all the particles were brought into the central region, guaranteeing their accessibility by the microgripper. Then they were positioned in a line of single beads by switching the frequency to 2.08 MHz. At this frequency—which also features one trapping plane along the centerline—the acoustic force field gradient was steeper, thus leading to better positioning accuracy. Then the microgripper was inserted parallel to the channel bottom surface and moved to the position of the first particle in the row. Once grasped, the particle was removed from the fluid and released on a glass plate. The water layer surrounding the particle proved to be sufficient to guarantee adhesion of the particles to the substrate rather than to the microgripper fingers. By repeating the process for the successive particles, the authors demonstrated that a position accuracy of 14 µm could be reached (measured when the microgripper was in position to grasp the particle but prior to the execution of this action), so that it was not necessary to reposition the microgripper in the lateral direction prior to grasping the next particle. Only a short activation of the ultrasound—usually turned off after the alignment—was sometimes necessary in order to reposition the particles on the centerline, which might have been displaced when the microgripper was moved in the channel or due to the fluid meniscus at the interface, which was slightly deformed when the microgripper was removed. In a further development of the system reported by Oberti et al. (2008), two side channels were added close to the entrance of the main channel. Through them, fluid was removed by means of a syringe pump. This way, particles could be brought toward the entrance of the channel, so that it was no longer necessary for the microgripper to travel a longer way after each removal. Two operation methods were described. The first consisted of prealigning the particles when they entered the channel and then—after having switched off the acoustic field—move them along the channel. This method had the disadvantage that particles traveling at different speeds or particles stuck on the substrate caused disruption of the sequence and loss of positioning. In the other method, particles were displaced while the ultrasonic field was on. In this case, as was seen in Figure 5.10, the force field caused particles to be retained toward the center of the channel, and thus the force generated by the pump had to overcome it. Nevertheless, both methods achieved a certain degree of automation.

Continuing along this direction, Oberti et al. (2009a) presented in 2008 a device for the preparation of protein crystals samples for x-ray crystallographic analysis. The idea was not to replace the standard method, but rather to use the same tools while facilitating the manipulation of the highly fragile crystals. Thus, instead of removing crystals from the well where they have been growing in a solution by "fishing" them out one by one using a special nylon loop mounted on a metal holder (a standard tool used for such applications), crystals of sizes ranging from 20 to 200 μm and shapes varying from diamond-like to cuboid-like were injected into a channel (cross section 1000 × 200 μm); see Figure 5.14. There, they were positioned in a row along the centerline and pulled toward an orifice etched into the lower channel surface (300 × 700 μm, oval shaped), by applying a fluid flow (0.05 mL s^{-1}) along the channel by means of a syringe pump connected to the opposite end of the device and operated in withdrawal mode. The orifice defined the liquid–air interface (leaking was prevented by surface tension). Once a crystal fell into the orifice, the loop was

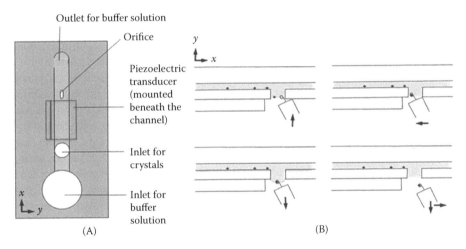

FIGURE 5.14 Schematics of the device by Oberti et al. (2009a) for the manipulation of protein crystals (not to scale). (A) Top view of the device consisting of a 1 mm wide, 200 μm deep channel etched in silicon and sealed on the top by a 1 mm thick glass plate. For operation, buffer solution was loaded into a reservoir, while crystals were injected by means of a pipette through a dedicated inlet. The acoustic field was generated by a piezoelectric transducer mounted beneath the channel and featuring the strip electrode configuration described by Neild et al. (2007a). Crystals were transported by applying a flow using a syringe pump through the area of highest acoustic force field, where they are positioned along the centerline. Then they are transported into the region close to an orifice defined on the bottom surface (oval shape, 300 × 700 μm). Once a crystal fell into the orifice, a nylon loop mounted on a metal holder was inserted from below, moved over the crystal, and then pushed down, until the surface tension force of the buffer solution meniscus was overcome. At the end, the crystal remained trapped in a droplet in the loop and could be used for further analysis. The sequence of crystal removal using the nylon loop is shown in (B).

inserted from below into the orifice at the opposite side where the crystal, was; it was moved above the crystal, which was removed by pushing it down until the surface tension was overcome. Once removed from the channel, the crystal, trapped in the droplet delimited by the loop, was flash-frozen to be used for further analysis.

5.4.7 PARTICLE CONCENTRATION WITHIN DROPLETS

The previous sections focused on acoustic fields excited within channels or chambers. Although some devices featured liquid–air interfaces (Neild et al., 2006a; Oberti et al., 2009a), they still consisted of cavities typically micromachined in silicon substrates. The potential of using small "open" volumes—usually a droplet placed on a surface—has recently been identified (see, for instance [Delamarche et al., 2005]). Not only can they be used as reaction volumes for biochemical processes, but these droplets can also be moved on the surface and merged together. Wixforth et al. (2004) proved that the use of surface acoustic waves causes acoustic streaming and leads to a net movement in the direction of the wave. In this section, the idea of using standing wave pressure fields for concentrating particles at certain locations will be discussed. Applications might be found in the enhancement of the binding of particles on a functionalized surface or in the collection of particles after a chemical reaction has occurred in order to remove them.

Oberti et al. (2009b) reported on the concentration of particles within a droplet (approximate diameter 3 mm) deposited on a 1 mm thick glass plate coated with a hydrophobic layer and mounted on a glass frame. On the plate reverse side, a square 4×4 mm piezoelectric transducer (0.5 mm in thickness) was attached and presented the same electrode configuration as previously used by the authors for the excitation of two-dimensional displacement fields (Oberti, 2009). The concentration of particles in a single small location, that is, a clump at the center of the droplet, was obtained by exploiting the same method as described in Section 5.3.3.1, consisting of exciting two perpendicular electrodes individually but alternately. First—when a single electrode was activated—a line was formed; then the signal was switched to the orthogonal electrode, and a line was formed approximately parallel to the other electrode. Already this line was shorter than the first line formed. After repeating the switching twice more, a clump of particles at the centre of the droplet was formed.

Concentration of particles at the center of a droplet was reported by Li et al. (2007) as well, using SAW generated by an interdigitated transducer (IDT) patterned on a lithium niobate substrate. Acoustic streaming generated in the droplet caused a shear-migration of particles convecting them toward the center of the droplet, forming an aggregate. The concentration rate was a function of the power and the particle size. For the typical operation parameters, aggregation of 45 μm diameter particles occurred in only 2 s, whereas for 10 μm diameter ones it took 15 s. The cell viability did not seem to be affected by the SAW radiation.

As a final remark, the recent work of Whitehill et al. (2010) is worth mentioning, in which gathered particles with a diameter of 10 to 116 μm were gathered in concentric rings and disks within a droplet (5 mm radius) by vibrating the surface onto which they were deposited at low frequency (60.59 and 110 Hz).

5.5   COMBINED MANIPULATION TECHNIQUES

Due to their different action ranges, acoustic manipulation can be combined with other contactless manipulation methods. The possibilities of simultaneous parallel handling of large numbers of particles offered by the long-range acting standing pressure fields are supplemented by the benefits of the localized, more accurate manipulation using electric, light, or magnetic fields. In the literature, the combination of acoustic manipulation with dielectrophoresis (DEP) and with magnetic manipulation has been reported.

The first study was published in 2006 by Wiklund et al. (2006). The system consisted of a 750 μm wide, 40 μm thick, 10 mm long channel defined by a silicon spacer between two glass plates. By means of three electrodes patterned obliquely with respect to the flow direction on its lower surface for DEP manipulation and a wedge transducer for acoustic manipulation, different operations could be performed. Single particles could be trapped in determined locations, separated according to their size (as large particles were less affected by the DEP force), and moved from one line into the next, when they reached the location of the electrodes.

In 2008, Ravula et al. (2008) described a device consisting of two 750 μm wide, 250 μm deep, 10 mm long glass channels bonded on a silicon or glass substrate, with a piezoelectric plate on the reverse side and an interdigitated electrode stripe patterned inside the channels on half of the bottom surface area. By the use of ultrasound (at 3 MHz, corresponding to operation at $3\lambda/2$, three lines would be formed) 10 μm diameter particles were preconcentrated in two parallel lines and then positioned more accurately in a chain of single particles exploiting negative dielectrophoresis, which placed them in a levitated state between two electrodes. The ability to focus particles is vital in flow cytometry, where such a system might find application. Furthermore, by changing to positive dielectrophoresis, the flowing particles could be retained against the flow in the channel. The authors suggested that by extending this idea and adding a perpendicular channel, it would be possible to trap particles at the intersection using electrical fields and flush them with chemicals.

Finally, a device combining acoustic and magnetic forces for the separation of mixtures of particles has been recently introduced by Adams et al. (2009). A mixture of magnetic particles (4.9 μm in diameter) and nonmagnetic polystyrene beads of two different sizes (1 and 5 μm in diameter) were separated along a 750 μm wide silicon channel. The particles flowing along a channel wall (this was achieved by injecting buffer from a second inlet and thereby constrain the particles to flow along one channel wall) were first exposed to a $3\lambda/2$ acoustic field excited at 2 MHz. The 1 μm particles, being too small to be deflected toward the closest of the two pressure nodal planes formed along the channel, continued undisturbed and left the channel through the first outlet. The other particles were instead moved toward the trapping plane. Close to the other extremity of the channel, the magnetic particles were forced by three permanent magnets on the bottom of the channel and pushed toward the wall opposite to that along which they were flowing when they entered the channel, by a series of patterned microfabricated ferromagnetic structures. These structures were obliquely orientated with an angle of 5°. This way, nonmagnetic particles left the main channel through an outlet located along the pressure nodal plane, and the magnetic ones were removed close to the wall. In order to reduce the amplitude of the acoustic radiation force acting on the magnetic particles in the last section of the channel, the channel width was reduced, in order to disrupt the resonant condition. In the reported experiment, 94.8% of magnetic beads were counted in the magnetic outlet, against 0.2% of nonmagnetic 5 μm diameter beads and 5.0% of the 1 μm diameter ones.

5.6 VIABILITY

Several systems and applications have been discussed in the previous sections, without addressing the issue of the viability of biological material upon exposure to ultrasound. In this section, some experimental results—for both macro- and microscopic systems—are described, confirming that, in general, no detrimental effect was observed. Viability is a function of the frequency: at lower frequencies, cavitation might cause damage to cells. However, cavitation is in general higher at the pressure antinodes; thus, cells being usually collected at the pressure nodes are less affected by it.

The viability of yeast cells in both standing and propagating waves at 2 MHz was tested by Radel et al. (2000b). The study reported that for the first case, no significant change could be observed. Morphology changes as well as a reduction of the duplication rate were revealed, instead. Moreover, it was observed that cells with degraded viability were no longer found at the pressure nodes.

Haake (2004) investigated the viability of Mesenchymal stem cells and Hela cells exposed to ultrasound generated by the previously described vibrating glass plate, using trypan blue dye to stain the cells. No significant influence of ultrasound could be observed over a total duration of 160 s and 25 min at 1.2 MHz, respectively. Radel et al. (2000b) observed that dead cells did not gather at the pressure nodes, as the living cells did, and attributed this to an alteration of the material parameters of the cells after death, which became similar to that of a fluid and thus caused a decrease of the acoustic radiation force. A similar observation was also made later by Oberti (2009) for micromachined systems.

The effect on two-dimensional aggregates of mammalian neural cells has been investigated by Bazou et al. (2005) under varying environmental conditions in a stainless steel chamber where a half-wavelength field was set up. First, a 1.5 MHz excitation frequency was used to excite a pressure amplitude at 0.54 MPa for 30 s, then it was reduced to 0.7 MPa for 2 min, and finally to 0.06 MPa for 1 h. The effect of several parameters was monitored. In particular, in relation to the temperature in the chamber, an increment of 0.5°C was observed over the first 30 min and then 0.005°C during long-term exposure. Furthermore, it was pointed out that intercellular forces due to acoustic fields did not exceed van der Waals forces during the test.

Moving to the field of micromachined systems, the study by Hultstrom et al. (2007) aimed at investigating the viability of adherent cells COS-7 trapped in a levitated state is noteworthy. They used a half-wavelength resonator, elliptically shaped in the plane, made of two glass plates spaced by a 260 μm thick PDMS layer and excited by a piezoelectric disk attached to the top of the device. The proliferation rate of recultured cells after exposure to ultrasound was measured. This method had the advantage of taking into account indirect effects in the cells, which appear only later in time. It was noted that the proliferation rate of exposed cells increased with respect to nonexposed ones, pointing out a beneficial effect of ultrasound on cell proliferation. This was attributed to a higher cell density for the exposed sample.

Evander et al. (2007) trapped and perfused yeast cells (1 μL/min with cell medium to promote their growth) for 6 h in such a 600 μm wide, 61 μm deep channel, operated at 1.24 MHz. Given the fact that growth was successfully achieved, damaging effects due to ultrasound were excluded by the authors. Furthermore, the viability of neural stem cells HiB5-GFP after 15 min of trapping was positively tested, albeit at a temperature increase of 7.2°C.

A final remark on temperature control needs to be made. Especially in systems operated in static fluid conditions and used for prolonged observation of cells, while they are trapped in the fluidic cavity over longer periods of time, a constant temperature is important in order to maintain biocompability without affecting the viability of cells. In this regard, Svennebring et al. (2007) focused their attention on temperature regulation in a chip of the same design as the one by Wiklund et al. (2006). To this end, they extended the system with a thermocouple, inserted 3 mm downstream of the wedge transducer. Different scenarios were tested, both for batch and flow-through operation (0 to 500 μL min^{-1}), with and without external temperature control. In static fluid conditions at constant voltage, measurements were performed over 12 h in order to determine the stability and repeatability of the temperature in the channel, which revealed a standard deviation of 0.3°C. Moreover, it was found out that in the interval 2–10 V_{pp} (voltages typically used in that system for the manipulation of cells), the temperature was independent of the flow rate. Furthermore, an accuracy of 0.1°C could be obtained by controlling the temperature, allowing an independent choice of the operating voltage. Finally, it must not to be forgotten that temperature control helps maintain a constant resonance frequency, thus obviating the need for complex electronic feedback systems.

5.7 CONCLUSIONS

The field of ultrasonic particle manipulation is in a fascinating state right now. Many developments have come together: Modeling of the whole system in three dimensions makes it feasible to predict the acoustic field in a cavity. Modeling of the interaction of this field with arbitrarily shaped particles and cells in general situations is still in its infancy, but is progressing rapidly. In terms of manufacturing, gradually a toolbox is being established that allows the use of ultrasound for specific applications. Many new avenues are open in this respect, from single-cell investigations, to studying the interaction of a small number of cells with each other or to determining the effects of various procedures on large ensembles of cells and particles. And last but not least, many technologies are available to characterize the systems, from an electromechanical to a cell biological point of view. The community is growing fast, which will further speed up the progress in many new directions. (See, e.g., USWnet, a network of scientists working in the field: www.uswnet.org)

REFERENCES

Adams, J. D., Thevoz, P., Bruus, H., and Soh, H. T. (2009). Integrated acoustic and magnetic separation in microfluidic channels. *Applied Physics Letters*, 95.

Augustsson, P., Åberg, L. B., Swärd-Nilsson, A. M. K., and Laurell, T. (2009). Buffer medium exchange in continuous cell and particle streams using ultrasonic standing wave focusing. *Microchimica Acta*, 164, 269–277.

Baker, N. V. (1972). Segregation and Sedimentation of Red Blood Cells in Ultrasonic Standing Waves. *Nature*, 239, 398–399.

Bazou, D., Kuznetsova, L. A., and Coakley, W. T. (2005). Physical environment of 2-D animal cell aggregates formed in a short pathlength ultrasound standing wave trap. *Ultrasound in Medicine and Biology*, 31, 423–430.

Bjerknes, V. (1909). *Die Kraftfelder*, Braunschweig, F. Vieweg.

Chladni, E. (1787). *Entdeckungen über die Theorie des Klanges*, Leipzig, Bey Weidmanns erben und Reich.

Coakley, W. T., Bardsley, D. W., Grundy, M. A., Zamani, F., and Clarke, D. J. (1989). Cell manipulation in ultrasonic standing wave fields. *Journal of Chemical Technology and Biotechnology*, 44, 43–62.

Danilov, S. D. and Mironov, M. A. (2000). Mean force on a small sphere in a sound field in a viscous fluid. *Journal of the Acoustical Society of America*, 107, 143–153.

Delamarche, E., Juncker, D., and Schmid, H. (2005). Microfluidics for processing surfaces and miniaturizing biological assays. *Advanced Materials*, 17, 2911–2933.

Doinikov, A. A. (1994). Acoustic radiation pressure on a compressible sphere in a viscous-fluid. *Journal of Fluid Mechanics*, 267, 1–21.

Dougherty, G. M., and Pisano, A. P. (2003). Ultrasonic particle manipulation in microchannels using phased co-planar transducers. 12th International Conference on Solid-State Sensors Actuators and Microsystems. *Boston Transducers'03: Digest of Technical Papers, Vols 1 and 2*, 670–673

Evander, M., Johansson, L., Lilliehorn, T., Piskur, J., Lindvall, M., Johansson, S., Almqvist, M., Laurell, T., and Nilsson, J. (2007). Noninvasive acoustic cell trapping in a microfluidic perfusion system for online bioassays. *Analytical Chemistry*, 79, 2984–2991.

Evander, M., Lenshof, A., Laurell, T., and Nilsson, J. (2008). Acoustophoresis in wet-etched glass chips. *Analytical Chemistry*, 80, 5178–5185.

Gherardini, L., Cousins, C. M., Hawkes, J. J., Spengler, J., Radel, S., Lawler, H., Devcic-Kuhar, B., and Groschl, M. (2005). A new immobilisation method to arrange particles in a gel matrix by ultrasound standing waves. *Ultrasound in Medicine and Biology*, 31, 261–272.

Gherardini, L., Radel, S., Sielemann, S., Doblhoff-Dier, O., Groschl, M., Benes, E., and Mcloughlin, A. J. (2001). A study of the spatial organisation of microbial cells in a gel matrix subjected to treatment with ultrasound standing waves. *Bioseparation*, 10, 153–162.

Glynne-Jones, P., Boltryk, R. J., Hill, M., Zhang, F., Dong, L., Wilkinson, J. S., Brown, T., Melvin, T., and Harris, N. R. (2010). Multi-modal particle manipulator to enhance bead-based bioassays. *Ultrasonics*, 50, 235–239.

Glynne-Jones, P., Hill, M., Harris, N. R., Townsend, R. J., and Ravula, S. K. (2008). The design and modeling of a lateral acoustic particle manipulator exhibiting quarter-wave operation. *Acoustics08*. Paris.

Gonzalez, I., Fernandez, L. J., Gomez, T. E., Berganzo, J., Soto, J. L., and Carrato, A. (2010). A polymeric chip for micromanipulation and particle sorting by ultrasounds based on a multilayer configuration. *Sensors and Actuators B-Chemical*, 144, 310–317.

Gorkov, L. P. (1961). Forces acting on a small particle in an acoustic field within an ideal fluid. *Doklady Akademii Nauk Sssr*, 140, 88–92.

Haake, A. (2004). Micromanipulation of Small Particles with Ultrasound. Zurich, Swiss Federal Institute of Technology, ETH.

Haake, A., and Dual, J. (2004). Positioning of small particles by an ultrasound field excited by surface waves. *Ultrasonics*, 42, 75–80.

Hagsater, S. M., Lenshof, A., Skafte-pedersen, P., Kutter, J. P., Laurell, T., and Bruus, H. (2008). Acoustic resonances in straight micro channels: Beyond the 1D-approximation. *Lab on a Chip*, 8, 1178–1184.

Harris, N., Hill, M., Shen, Y., Townsend, R. J., Beeby, S., and White, N. (2004). A dual frequency, ultrasonic, microengineered particle manipulator. *Ultrasonics*, 42, 139–144.

Harris, N. R., Hill, M., Beeby, S., Shen, Y., White, N. M., Hawkes, J. J., and Coakley, W. T. (2003). A silicon microfluidic ultrasonic separator. *Sensors and Actuators B-Chemical*, 95, 425–434.

Hawkes, J. J., Barber, R. W., Emerson, D. R., and Coakley, W. T. (2004a). Continuous cell washing and mixing driven by an ultrasound standing wave within a microfluidic channel. *Lab on a Chip*, 4, 446–452.

Hawkes, J. J., and Coakley, W. T. (2001). Force field particle filter, combining ultrasound standing waves and laminar flow. *Sensors and Actuators B-Chemical*, 75, 213–222.

Hawkes, J. J., Long, M. J., Coakley, W. T., and McDonnell, M. B. (2004b). Ultrasonic deposition of cells on a surface. *Biosensors and Bioelectronics*, 19, 1021–1028.

Hill, M., Townsend, R. J., and Harris, N. R. (2008). Modelling for the robust design of layered resonators for ultrasonic particle manipulation. *Ultrasonics*, 48, 521–528.

Hultstrom, J., Manneberg, O., Dopf, K., Hertz, H. M., Brismar, H., and Wiklund, M. (2007). Proliferation and viability of adherent cells manipulated by standing-wave ultrasound in a microfluidic chip. *Ultrasound in Medicine and Biology*, 33, 145–151.

Johansson, L., Nilsson, M., Lilliehorn, T., Almqvist, M., Nilsson, J., Laurell, T., and Johansson, S. (2005). An evaluation of the temperature increase from PZT micro-transducers for acoustic trapping. *2005 IEEE Ultrasonics Symposium, Vols 1–4*, 1614–1617.

Jonsson, H., Holm, C., Nilsson, A., Petersson, F., Johnsson, P., and Laurell, T. (2004). Particle separation using ultrasound can radically reduce embolic load to brain after cardiac surgery. *Annals of Thoracic Surgery*, 78, 1572–1578.

Kapishnikov, S., Kantsler, V., and Steinberg, V. (2006). Continuous particle size separation and size sorting using ultrasound in a microchannel. *Journal of Statistical Mechanics—Theory and Experiment*, 2006, P01012.

King, L. V. (1934). On the Acoustic Radiation Pressure on Spheres. *Proceedings of the Royal Society of London. Series A, Mathematical and Physical Sciences*, 147, 212–240.

Kinsler, L. E., Frey, A. R., Coppens, A. B., and Sanders, J. V. (1982). *Fundamentals of Acoustics*. New York, Wiley.

Kundt, A. (1866). Ueber eine neue Art akustischer Staubfiguren und über die Anwendung derselben zur Bestimmung der Schallgeschwindigkeit in festen Körpern und Gase. *Annalen der Physik*, 127, 947–523.

Kuznetsova, L. A. and Coakley, W. T. (2007). Applications of ultrasound streaming and radiation force in biosensors. *Biosensors & Bioelectronics*, 22, 1567–1577.

Laurell, T., Petersson, F., and Nilsson, A. (2007). Chip integrated strategies for acoustic separation and manipulation of cells and particles. *Chemical Society Reviews*, 36, 492–506.

Lenshof, A., Ahmad-Tajudin, A., Jaras, K., Sward-Nilsson, A. M., Aberg, L., Marko-Varga, G., Malm, J., Lilja, H., and Laurell, T. (2009). Acoustic whole blood plasmapheresis chip for prostate specific antigen microarray diagnostics. *Analytical Chemistry*, 81, 6030–6037.

Li, H., Friend, J. R., and Yeo, L. Y. (2007). Surface acoustic wave concentration of particle and bioparticle suspensions. *Biomedical Microdevices*, 9, 647–656.

Li, H. Y., Friend, J., Yeo, L., DasvarmA, A., and Traianedes, K. (2009). Effect of surface acoustic waves on the viability, proliferation and differentiation of primary osteoblast-like cells. *Biomicrofluidics*, 3, 034102.

Lilliehorn, T., Nilsson, M., Simu, U., Johansson, S., Almqvist, M., Nilsson, J., and Laurell, T. (2005a). Dynamic arraying of microbeads for bioassays in microfluidic channels. *Sensors and Actuators B-Chemical*, 106, 851–858.

Lilliehorn, T., Simu, U., Nilsson, M., Almqvist, M., Stepinski, T., Laurell, T., Nilsson, J., and Johansson, S. (2005b). Trapping of microparticles in the near field of an ultrasonic transducer. *Ultrasonics*, 43, 293–303.

Manneberg, O., Hagsater, S. M., Svennebring, J., Hertz, H. M., Kutter, J. P., Bruus, H., and Wiklund, M. (2009a). Spatial confinement of ultrasonic force fields in microfluidic channels. *Ultrasonics*, 49, 112–119.

Manneberg, O., Svennebring, J., Hertz, H. M., and Wiklund, M. (2008a). Wedge transducer design for two-dimensional ultrasonic manipulation in a microfluidic chip. *Journal of Micromechanics and Microengineering*, 18.

Manneberg, O., Vanherberghen, B., Onfelt, B., and Wiklund, M. (2009b). Flow-free transport of cells in microchannels by frequency-modulated ultrasound. *Lab on a Chip*, 9, 833–837.

Manneberg, O., Vanherberghen, B., Svennebring, J., Hertz, H. M., Onfelt, B., and Wiklund, M. (2008b). A three-dimensional ultrasonic cage for characterization of individual cells. *Applied Physics Letters*, 93(6) 063901.

Martin, S. P., Townsend, R. J., Kuznetsova, L. A., Borthwick, K. A. J., Hill, M., McDonnell, M. B., and Coakley, W. T. (2005). Spore and micro-particle capture on an immunosensor surface in an ultrasound standing wave system. *Biosensors and Bioelectronics*, 21, 758–767.

Neild, A., Oberti, S., Beyeler, F., Dual, J., and Nelson, B. J. (2006a). A micro-particle positioning technique combining an ultrasonic manipulator and a microgripper. *Journal of Micromechanics and Microengineering*, 16, 1562–1570.

Neild, A., Oberti, S., and Dual, J. (2007a). Design, modeling and characterization of microfluidic devices for ultrasonic manipulation. *Sensors and Actuators B-Chemical*, 121, 452–461.

Neild, A., Oberti, S., Radziwill, G., and Dual, J. (2007b). Simultaneous positioning of cells into two-dimensional arrays using ultrasound. *Biotechnology and Bioengineering*, 97, 1335–1339.

Neild, A., Oberti, S., Haake, A., and Dual, J. (2006b). Finite element modeling of a microparticle manipulator. *Ultrasonics*, 44, E455–E460.

Nilsson, A., Petersson, F., Jonsson, H., and Laurell, T. (2004). Acoustic control of suspended particles in micro fluidic chips. *Lab on a Chip*, 4, 131–135.

Norris, J. V., Evander, M., Horsman-Hall, K. M., Nilsson, J., Laurell, T., and Landers, J. P. (2009). Acoustic Differential Extraction for Forensic Analysis of Sexual Assault Evidence. *Analytical Chemistry*, 81, 6089–6095.

Oberti, S. (2009). Micromanipulation of small particles within micromachined fluidic systems using ultrasound. Zurich, Swiss Federal Institute of Technology, ETH.

Oberti, S., Möller, D., Gutmann, S., Neild, A., and Dual, J. (2009a). Novel sample preparation technique for protein crystal x-ray crystallographic analysis combining microfluidics and acoustic manipulation. *Journal of Applied Crystallography*, 42, 636–641.

Oberti, S., Neild, A., Quach, R., and Dual, J. (2009b). The use of acoustic radiation forces to position particles within fluid droplets. *Ultrasonics*, 49, 47–52.

Oberti, S., Neild, A., and Dual, J. (2007). Manipulation of micrometer sized particles within a micromachined fluidic device to form two-dimensional patterns using ultrasound. *Journal of the Acoustical Society of America*, 121, 778–785.

Oberti, S., Neild, A., Möller, D., and Dual, J. (2008). Towards the automation of micron-sized particle handling by use of acoustic manipulation assisted by microfluidics. *Ultrasonics*, 48, 529–536.

Petersson, F., Aberg, L., Sward-Nilsson, A. M., and Laurell, T. (2007). Free flow acoustophoresis: Microfluidic-based mode of particle and cell separation. *Analytical Chemistry*, 79, 5117–5123.

Petersson, F., Nilsson, A., Holm, C., Jonsson, H., and Laurell, T. (2005a). Continuous separation of lipid particles from erythrocytes by means of laminar flow and acoustic standing wave forces. *Lab on a Chip*, 5, 20–22.

Petersson, F., Nilsson, A., Jonsson, H., and Laurell, T. (2005b). Carrier medium exchange through ultrasonic particle switching in microfluidic channels. *Analytical Chemistry*, 77, 1216–1221.

Radel, S., Gherardini, L., Mcloughlin, A. J., Doblhoff-Dier, O., and Benes, E. (2000a). Breakdown of immobilisation/separation and morphology changes of yeast suspended in water-rich ethanol mixtures exposed to ultrasonic plane standing waves. *Bioseparation*, 9, 369–377.

Radel, S., Mcloughlin, A. J., Gherardini, L., Doblhoff-Dier, O., and Benes, E. (2000b). Viability of yeast cells in well controlled propagating and standing ultrasonic plane waves. *Ultrasonics*, 38, 633–637.

Ravula, S. K., Branch, D. W., James, C. D., Townsend, R. J., Hill, M., Kaduchak, G., Ward, M., and Brener, I. (2008). A microfluidic system combining acoustic and dielectrophoretic particle preconcentration and focusing. *Sensors and Actuators B-Chemical*, 130, 645–652.

Ristic, V. M. (1983). *Principles of Acoustic Devices*, New York, Wiley.

Saito, M., Kitamura, N., and Terauchi, M. (2002). Ultrasonic manipulation of locomotive microorganisms and evaluation of their activity. *Journal of Applied Physics*, 92, 7581–7586.

Schwarz, T., and Dual, J. (2009). Rotation of non spherical particles with amplitude modulation. *USWNet*. Stockholm.

Svennebring, J., Manneberg, O., Skafte-Pedersen, P., Bruus, H., and Wiklund, M. (2009). Selective bioparticle retention and characterization in a chip-integrated confocal ultrasonic cavity. *Biotechnology and Bioengineering*, 103, 323–328.

Svennebring, J., Manneberg, O., and Wiklund, M. (2007). Temperature regulation during ultrasonic manipulation for long-term cell handling in a microfluidic chip. *Journal of Micromechanics and Microengineering*, 17, 2469–2474.

Townsend, R. J., Hill, M., Harris, N. R., and McDonnell, M. B. (2008). Performance of a quarter-wavelength particle concentrator. *Ultrasonics*, 48, 515–520.

Townsend, R. J., Hill, M., Harris, N. R., and White, N. M. (2006). Investigation of two-dimensional acoustic resonant modes in a particle separator. *Ultrasonics*, 44, E467–E471.

Wang, J. T., and Dual, J. (2009). Numerical simulations for the time-averaged acoustic forces acting on rigid cylinders in ideal and viscous fluids. *Journal of Physics A—Mathematical and Theoretical*, 42, 17.

Whitehill, J., Neild, A., Ng, T. W., and Stokes, M. (2010). Collection of suspended particles in a drop using low frequency vibration. *Applied Physics Letters*, 96. 053501

Wiklund, M., Gunther, C., Lemor, R., Jager, M., Fuhr, G., and Hertz, H. M. (2006). Ultrasonic standing wave manipulation technology integrated into a dielectrophoretic chip. *Lab on a Chip*, 6, 1537–1544.

Wixforth, A., Strobl, C., Gauer, C., Toegl, A., Scriba, J., and Von Guttenberg, Z. (2004). Acoustic manipulation of small droplets. *Analytical and Bioanalytical Chemistry*, 379, 982–991.

Yosioka, K. (1955). Acoustic radiation pressure on a compressible sphere. *Acustica* 5, 167–173.

6 3D Biomanipulation Using Microgrippers

Peter Bøggild
Department of Micro- and Nanotechnology,
Technical University of Denmark, Lyngby

CONTENTS

6.1 INTRODUCTION

In our "macroscale" world, physical manipulation of objects using tools has been the cornerstone of technological progress for thousands of years. For humans, the ability to grip—to apply force through independent movement of the opposing thumb—plays a key role in on our ability to manipulate quickly and precisely a vast range of different objects of different size, shape, and materials. Machines grip and release objects in factories everywhere; more than a million industrial robots are currently in use worldwide.

How about the microscale? The glass capillary invented by Marshall Barber (Korzh and Strahle, 2002) has for nearly a century served as a ubiquitous tool in biology, playing a key role in intracytoplasmic sperm injection (ICSI), DNA injection, gene therapy, and many biomedical applications (Desai et al., 2007). While the simplicity of this tool is one of its advantages, there is no question that development of better, gentler, more versatile, and more precise tools continue to have a large impact on what is possible in microbiology. It is, however, not at all clear to what extent tweezers, grippers, pliers, spoons, forks, not to mention hands, apply to structures that are far smaller than the macroscale, and perhaps even alive.

As a spin-out from the incredible success of silicon microfabrication technology and integrated circuits, the field of MEMS (Micro Electromechanical Systems) has created accurate, microscale mechanical actuators and sensors for thousands of consumer products. This is an obvious starting point for development of a mechanical microgripper, and a plethora of prototypic manipulation tools have indeed been reported since the beginning of the 1990s (Kim et al., 1992). Manipulation of biological samples is even more challenging than inorganic micro- and nanostructures, due to their greater softness and fragility. For this reason, microscale grippers and tweezers designed with biological manipulation in mind have not yet been used in practice except for a few "proof-of-principle" experiments.

The chapter will first consider the challenges and key concepts of gripper-based biomanipulation. As nanotweezers and nanogrippers originate from the "dry" MEMS field, it is instructive to review the state of art in mechanical 3D tweezers (Sahu et al., 2010), with a particular focus on the two most common types of gripper "motor," which use either electrothermal or electrostatic actuators. Finally, the chapter will discuss the feasibility of gripper-based nanomanipulation.

6.2 BASIC CONCEPTS OF GRIPPER MANIPULATION

As suggested earlier, many incarnations of microgrippers have some resemblance to human hands, and this familiarity makes them different from other types of micromanipulation. As humans, we are automatically experts in three-dimensional manipulation on the macroscale; we might benefit from this experience for developing strategies for aligning, picking, guiding, and placing objects, and for designing manipulation tools. However, this preloaded "know-how" can sometimes be an obstacle for developing different and more effective manipulation concepts. The rules, conditions, and constraints for manipulation of cells is very different from our daily experience, and we might be in for a surprise. We begin here by discussing a few basic notions of microgrippers: pick-and-place, actuators, end-effectors, force-feedback, and guiding.

6.2.1 Pick and Place

"Pick-and-place" systems are a well-known industrial assembly. In particular, the electronics industry has developed sophisticated automatic systems for assembly and bonding of surface-mounted electrical components onto printed circuit boards. To pick and place any object in an effective manner, some control of the interaction forces is required, which is particularly true for biological samples.

We can think of a "force balance" between the forces acting between the object and the surface $F_{\text{subject-object}}$ and the forces acting between the component and the tool, $F_{\text{object-tool}}$. Manipulation can then be done in either a passive or an active mode, depending on whether the tool itself actively changes $F_{\text{object-tool}}$ during the process. With a *passive* tool (i.e., an STM tip), you may arrange the interaction forces to promote picking up (object preferring contact with tool rather than the substrate), $F_{\text{surface1-object}} < F_{\text{object-tool}}$, and subsequently promote placing (object prefers contact with the target over the tool, $F_{\text{surface2-object}} < F_{\text{object-tool}}$). This could be done by choosing a source surface material with a low adhesion, a material for the tool with intermediate adhesion and a relatively high adhesion target. This is not in fact a particularly simple or practical approach. Haliyo and co-workers demonstrated picking of microstructures with a cantilever just by van der Waals forces (Haliyo et al., 2002), but implemented release by fast acceleration of the particle by vibrating the cantilever, essentially shaking the microparticles off; as surface forces dominate over volumetric forces at low particle sizes, such a strategy becomes increasingly difficult for smaller objects. The interaction between the tool and the object can also be reduced by coating the tool with a low-surface-energy film (such as Teflon), or by increasing the nanoscale roughness (Tam et al., 2009; Weck and Peschke, 2004).

Dynamical tools such as grippers and tweezers can be used to vary the interaction force between object and tool in a way similar to the human hand, through a variable mechanical force $F_{\text{grip}}(V)$, which is controlled by some external parameter V; this could be an electrical bias voltage, bias current, temperature, magnetic field, or something entirely different used to control the actuation. To release, the gripping force is set to zero, $F_{\text{grip}}(V) = 0$. Likewise, gluing or soldering can be used to locally enhance the adhesion between the object and the target surface, and thereby help finally to release the object in the desired position. In an aqueous solution, adhesion forces are generally smaller; capillary forces are eliminated, and van der Waals forces are strongly diminished (Bhushan, 2010; Fernandez-Varea and Garcia-Molina, 2000). This makes adhesion-based manipulation less effective, but also reduces the problem of stiction of objects to the gripper. The force balance for picking and placing, respectively, might then look like this:

$$\text{Pick:} \quad F_{\text{source-object}} < F_{\text{object-tool}} + F_{\text{grip}}(V)$$

$$\text{Place:} \quad F_{\text{target-object}} + F_{\text{glue}} > F_{\text{object-tool}}$$

(6.1)

Here, the variable gripping force can be used to overcome the initial adhesion forces, $F_{\text{subject-object}}$, while the "glue" force, F_{glue}, helps to overcome the adhesion between the tool and the object.

Figure 6.1 shows in a schematic form a generic sequence of steps involved in pick-and-place, from the location and identification of the object to the final placement. First, the object must located and identified, which can be a serious challenge, since the objects are small, the surroundings are often featureless (e.g., a glass-slide), and the image quality can be very poor. Then the tool must be aligned with the object. This is difficult because the microscope image shows a 2D projection of a 3D scene; it can be very hard to determine if the object is in the same plane as the tool. Also, detecting when the gripper/tool is in contact is difficult. The tool must then be closed

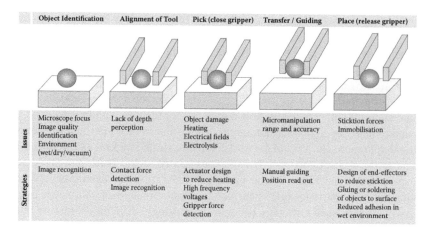

Object Identification	Alignment of Tool	Pick (close gripper)	Transfer / Guiding	Place (release gripper)

Issues

| Microscope focus
Image quality
Identification
Environment
(wet/dry/vacuum) | Lack of depth
perception | Object damage
Heating
Electrical fields
Electrolysis | Micromanipulation
range and accuracy | Sticktion forces
Immobilisation |

Strategies

| Image recognition | Contact force
detection
Image recognition | Actuator design
to reduce heating
High frequency
voltages
Gripper force
detection | Manual guiding
Position read out | Design of end-effectors
to reduce sticktion
Gluing or soldering
of objects to surface
Reduced adhesion in
wet environment |

FIGURE 6.1 Illustration of the generic manipulation sequence, the issues involved in the different steps, as well as which types of strategies can be considered in solving the issues.

around the object; here, the actuation principle and actuator design play a role; electrothermal actuation can, for instance, lead to heating of the object or the liquid surroundings, and lead to unwanted damage or changes. High voltages in a saline solution can lead to bubble formation through hydrolysis, which disrupts the process. To counteract this, the actuator has to be carefully designed to influence the object minimally. The object needs to be transferred to the target area, which, compared to the size of the object, can be a formidable distance, requiring accurate position control. Finally, the object is to be positioned and immobilized, and here surface forces can present a problem; the object might stick to the tool, requiring either special action to be taken such as vibrating the tool mechanically or gluing one end of the object to the surface, to overcome the interactions between the tool and the object.

6.2.2 ACTUATORS

Actuators are structures that convert some type of energy into motion (translation, rotation) (Krijnen et al., 2003), simultaneously playing the role of bones (structural elements) and muscles (moving elements) of microelectromechanical structures. In the following, we refer to actuation as the opening and closing mechanism, rather than the positioning of the gripper, although this is also done with actuators. This typically calls for compact, effective, integrated actuator principles. Electrostatic and electrothermal actuators are by far the most commonly used types for microgrippers and tweezers, as these are fully compatible with conventional silicon microfabrication (Molhave et al., 2004), and can be realized with fairly few fabrication steps, and are relatively compact. Piezoelectric actuation is fast and provides ultrahigh resolution, yet involves special ceramic materials that are nontrivial to integrate with silicon technology and often leads larger, more bulky structures (Chen et al., 2009). Piezoactuators are thus mostly used for micropositioning than opening/closing the grippers. These types of grippers all involve the application of electric voltages. To overcome the problems caused by high electrical fields and heat generated by

TABLE 6.1
Typical Characteristics

Characteristics	Electrothermal	Electrostatic
Force	10 µN–10 mN	10 µN–100 µN
Displacement	Large (20 µm)	Small to large (2–50 µm)
Voltage	Low (0–15 V)	Med (20–100V)
Bandwidth	Low (500 Hz)	High (kHz)
Footprint (size)	<<1 mm^2	>1 mm^2

Source: Adapted from Sahu, B., Taylor, C. R., and Leang, K. K. *Journal of Manufacturing Science and Engineering—Transactions of the ASME*, 132. 030917-3. 2010. With permission.

actuation devices, alternative actuators such as ionic conducting polymer films (Zhou et al., 2004) and pneumatic actuators (Lu and Kim, 2006) have been proposed.

Hubbard et al. (2006) overview the most common types of actuators for micro- and nanopositioning, and an excellent review of micro- and nanomanipulation techniques is given by Sahu et al., 2010. Typical characteristics for electrostatic and electrothermal actuators are shown in Table 6.1, which is adapted from works by Hubbard et al. (2006) and Sahu et al. (2010). In the following, the emphasis will be on electrostatic and electrothermal actuation.

6.2.3 END-EFFECTORS

The end-effector is the part of the gripper interacting with the sample. Since biological materials such as living cells have a great variety of mechanical properties, sizes, and shapes, it is impossible to define a generic end-effector that will work satisfactorily in all situations, unless it is capable of reshaping itself; it is often necessary to shape the end-effector in a certain way to allow a firm grip without destroying the sample. This is a similar situation to macroscale pliers and tweezers, which exist in hundreds of different shapes and sizes depending on their use. End-effectors are often realized as straight parallel beams of polymer or silicon, or as converging, chop-stick-like beams (Park and Moon, 2005). Chronis et al. (2005) use a rounded, concave end-effector to better fit the shape of spherical particles.

As the gripper fabrication process is often cumbersome, lengthy, and expensive, it is generally inconvenient to change the shape of the end-effectors more than is absolutely necessary. One approach to increasing the customizability is to finalize the shape of the end-effectors individually using fast prototyping. Sardan and co-workers (2009) used Focused Ion Beam milling to end-effectors of electrothermal grippers with 100 nm spatial resolution to reduce the contact area and thereby reduce adhesion effects (see Figure 6.2). In terms of biomanipulation, an issue here is biocompatibility. While materials such as silicon, silicon dioxide, as well as the polymer SU-8 (Chronis and Lee, 2005; Elbuken et al., 2008a) are biocompatible,

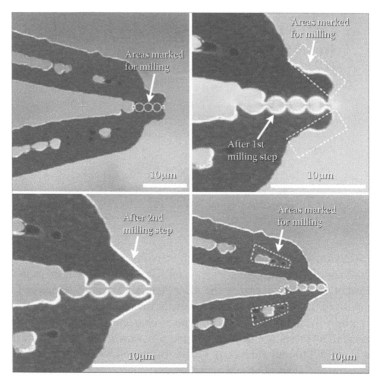

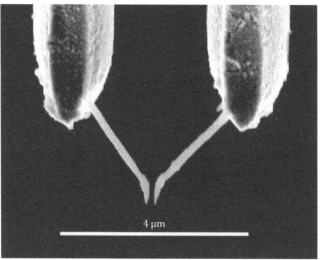

FIGURE 6.2 Top: Focused ion beam milling allows accurate shaping of the end-effectors to reduce the contact area, or adapt the shape to obtain a more firm or more gentle grip (Sardan, et al. *Focused Ion Beam (Fib) Modification Of Topology Optimized Polysilicon Microgrippers*, 2009; Courtesy of Özlem Sardan.) Bottom: a pair of nanotweezers, where carbonaceous end-effectors have been created through electron-beam-induced deposition, leaving a 100 nm gap in between. (Modified from Bøggild, et al., 12, 331–335, 2001.) Copyright 2001 IOP Publishing.

contamination by ions (such as Gallium) in the focused ion beam milling process is highly unwanted. In this case, a coating with a biocompatible film can become necessary (Duch et al., 2009).

An alternative technique is electron beam induced deposition, where carbonaceous or metal-containing molecules are decomposed by a focused electron beam; this process will lead to formation of long needles, which can be used to fine-tune the shape of the end-effectors. Such needles were used to fabricate chop-stick type nanotweezers with a gap size in the range 25–100 nm (Boggild et al., 2001).

6.2.4 FORCE FEEDBACK AND POSITION READOUT

Handling of fragile objects is safer and easier if the force applied to the objects can be adjusted to match their mechanical properties. This requires some form of force measurement to be integrated with the gripper. Force feedback can be done by measuring the actual position x_{actual} of one or two arms, and comparing to the expected position $x_0(V)$ at the specific bias voltage V. If the arm is displaced $\Delta x = x_{\text{actual}} - x_0(V)$ by contact with an object, and the spring constant k of the arm is already known, the applied gripper force can be calculated from Hooke's law:

$$F = -k(x_{\text{actual}} - x_0(V)) \tag{6.2}$$

The expected position is the free actuation corresponding to a single, isolated arm. We note that for a gripper with initial gap size g_0, each arm can only move to the half-gap distance $g_0/2$, at which it will be blocked by the other arm.

The position x_{actual} can be detected in different ways, of which capacitive and piezoresistive readout are by far the most commonly used in MEMS-based grippers:

1. **Capacitive.** The idea is to monitor the change of capacitance $C = \varepsilon A/g$ as either the distance g between two capacitor plates, or the area A, is varied, with ε being the electrical permittivity. One or two of the capacitor structures need to be connected mechanically to the gripper arm, so that the position of the gripper arm can be calculated from the change in capacitance. Very often, the actual layout of the capacitor plates is a set of interdigitated fingers, similar to the electrostatic comb-drives described in Section 6.3. Capacitive readout devices are typically bulky, but can give very accurate results. Kim and co-workers (Kim et al., 2008) realized a two-axis readout device capable of measuring the gripping force in the nanonewton regime. Umemoto et al. (2009) used a high-frequency self-oscillation of a comb-drive-based nanogripper with capacitive read out as a proximity contact detection method; when the gripper was just touching any object, the self-oscillation ceased. In this way, the authors achieved two types of force feedback with a single capacitive read-out sensor.
2. **Piezoresistive.** Most conducting materials change their electrical resistivity ρ upon mechanical strain, and in some cases the response is substantial. For a rod with length L, an elongation ΔL corresponds to the mechanical

strain $\Delta L/L$. The piezoresistive change of resistance $\Delta R/R$ can then be expressed by the gauge factor $g = (\Delta R/R)/(\Delta L/L)$, which is typically of order unity for metals and much higher for semiconductors. Doped silicon has a gauge factor in the range -130 (n-doped) to 200 (p-doped), with polycrystalline silicon having substantially smaller values (below 100). It is known that nanostructures such as nanowires (Reck et al., 2008) and carbon nanotubes (Tombler et al., 2000) can have very large gauge factors, well above 1000. By incorporating a piezoresistive sensor at or near the surface of a flexible structure, bending will be translated into deformation of the piezoresistor in well-defined manner. In this way, changes in the position for the end-effector can be calculated; piezoresistive cantilever readout is a well-established technique in Atomic Force Microscopy (AFM) and biosensing (Thaysen et al., 2000; Marie et al., 2005). It is fairly straightforward to integrate silicon resistors in the cantilevers and beams using silicon microfabrication. Unfortunately, piezoresistive readout is better suited to detect out-of-plane bending, whereas the gripper arms almost always deflect toward each other in the plane; see Figure 6.3. Integrating piezoresistive readout for lateral gripper force feedback is challenging, and requires creative solutions (Chen et al., 2009; Molhave and Hansen, 2005).

In Atomic Force Microscopy, optical readout is used to detect bending, that is, by directing a laser beam toward the cantilever and monitoring the deflection (Bhushan, 2010). Manipulation with AFM probes is described in Chapter 2. While being very powerful for when the gripper collides or touches something, this method is less obvious for gripper force measurement, as it favors out-of-plane rather than in-plane detection. Nevertheless, successful realization of lateral force feedback has been achieved on a gripper by directing the laser beam toward the side of the actuator (Zhou and Nelson, 1999), by tilting the readout system 90°.

6.2.5 Guiding and Tracking Inside a Microscope

For 3D gripper-based biomanipulation to become truly useful, effective guiding and positioning methods must be developed. The ability to quickly and precisely align the gripper jaws to the object depends on the visibility and accessibility defined by the environment, the microscope, and the arrangement of micromanipulators and grippers. In terms of resolution, environmental scanning electron microscopes (Stokes, 2008) allow resolutions below 10 nm in low-pressure water vapor conditions (typically up to 10 mbar) inside the sample chamber. Ahmad and co-workers (Ahmad et al., 2008) demonstrated penetration and mechanical investigations of the cell membranes of yeast cells using dual micro needles inside an ESEM, without destroying the cell. The requirements of a high microscope resolution combined with a large working distance makes effective perception of depth difficult. This is similar to nano- and micromanipulation in a vacuum environment, where sophisticated depth detection and image recognition methods have been developed for tracking, guiding, and aligning of nano-tools in 3D; see, for instance, work by Eichhorn and

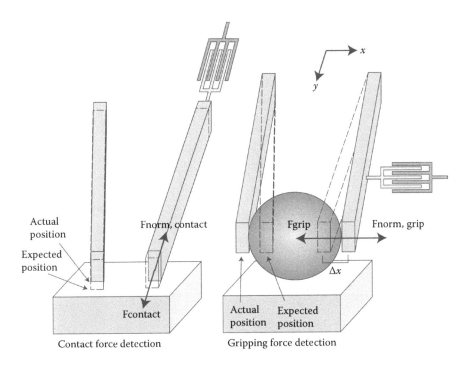

FIGURE 6.3 Schematic of contact force and gripping force detection using comb-shaped capacitive readout devices. If the gripper touches the surface, the pins in the *y*-directional sensor are pushed into each other, giving rise to a measurable change of capacitance. From these variations, the position, and eventually the contact force $F_{contact}$ in the *y*-direction, can be determined. Likewise, the gripping force can be estimated by measuring the offset Δx from the expected position, and by using the known spring constant in the *x*-direction. Both sensors can be integrated in the same gripper system if required. (Kim, K., Liu, X. Y., Zhang, Y., and Sun, Y. *Journal of Micromechanics and Microengineering*, 18, 2008.)

co-workers (2008). Scanning electron microscopes are in general characterized by a very long focal depth; this means that two objects, which both appear sharp in the view field, are not necessarily close to each other along the optical axis. This is very convenient for imaging, where several objects distributed in space can be observed at the same time. Alignment of, for instance, a gripper with another object can, however, be very difficult, since objects at different distances from the objective lens appear to be in the same plane. While scanning electron microscopes are sometimes used to characterize the morphology of biomaterials, the optical microscope is the more relevant option for live samples. In optical microscopy, the very limited focal depth compared to, for instance, electron microscopy, can actually be helpful; by focusing on the objects to be manipulated, and slowly bringing the tool into focus as well, the co-alignment of tool and object can be achieved by adjusting to the same plane, simply by keeping all objects of interest in the same narrow focal plane. If biological cells are located on a planar, transparent surface, aligning the focal plane with the sample plane is even easier.

(A)

(B)

(C)

(D)

FIGURE 6.4 (See color insert.) Examples of multiprobe systems for micro- and nanomanipulation. (A): Four micromanipulators from Klocke Nanotechnik (http://www.nanomotor. de/), (B): A 13-degrees of freedom nanomanipulator from Smaract (http://www.smaract.de/). (C): A multiprobe unit from Zyvex (http://www.zyvex.com/). (D): Piezoelectric micromanipulator from DTI (http://www.dti-nanotech.com/).

The quality and type of the micromanipulation system can be crucial. Apart from standard micromanipulators for patch clamp and microinjection offered by, for instance, by Nikon (http://www.nikon.com/), Burleigh (http://www.exfo-burleigh.com/) and others, there are several manipulator systems on the market for 3D multiprobe, high-precision micro- and nanomanipulation with sophisticated user interfaces, which might be adapted to biological manipulation. These include, for instance, Klocke Nanotechnik, who offer advanced high-precision multiprobe manipulation systems with the possibility of automation; Kleindieck Nanotechnik, who besides multiprobe systems also produce high-precision manipulators for life sciences; Smaract, who make a range of versatile, affordable single and multi-manipulators and an easy-to-use interface; and US-based Zyvex Instruments, who provide integrated solutions with multiprobe manipulators as well as manufacture their own electrothermal gripper. While the primary target for most of these systems appears to be the semiconductor industry, these are high-quality tools that have been optimized for reproducibility, reliability, resolution, and versatility, and should be considered a suitable basis for an advanced biomanipulation setup, in particular, if more advanced computer-controlled operations or even automation are needed.

In the vast majority of experiments with gripper-based manipulation reported to date, manual operation and aligning has been the only option. There has been considerable work on automated control systems for cell micromanipulation and microinjection (Kuncova et al., 2004; Huang et al., 2008; Xie et al., 2009; Zhang et al., 2007; Sakaki et al., 2009), yet in terms of automation or semiautomation of *grippers* for biological samples, there is a limited literature; see overview by Ouyang et al. (2007) . While the research field is developed, the technological aspect may still be regarded as being in its infancy; the problems regarding design of truly bio- and water-compatible gripper-inspired tools is still the primary concern. As these problems are solved, automation will be the main limiting factor for scaling this technology to practical application such as batch biomanipulation. While this research should serve as a starting point for development of automatic guiding and pick-and-place of 3D grippers, the literature of control systems for vacuum microrobotics should also be considered (Fatikow and Rembold, 2002; Fatikow et al., 2007; Fatikow et al., 2008; Weck and Peschke, 2004), as many of the challenges and solutions found here should overlap with those needed for effective gripper biomanipulation.

6.3 ELECTROSTATIC GRIPPERS

Electrostatic actuators are basically capacitors with air, liquid, or vacuum as a dielectric. When applying a voltage difference to the capacitor plates, the charge difference on the plates leads to a net electrostatic force—and this is used to drive the actuator. The electrostatic actuator is by far the most widely used in MEMS for resonators, switches, micromirrors, hard disk drives, accelerometers, sensors, and so forth. Since the first MEMS microgripper reported by Pisano and coworkers (Kim et al., 1992; Kim and Pisano, 1992), mainly two types of electrostatic actuators have been used for grippers and tweezers: the plate capacitor drive and the comb-drive; see Figures 6.5 and 6.7. Both types of drive exhibit a parabolic displacement as a function of voltage; however, the plate capacitor drive suffers from a so-called snap-in or pull-in instability, leading to a collapse of the plates when the gap reaches 2/3 of the neutral position. This occurs as the elastic restoration force of the mechanical actuator structure can no longer keep up with the electrostatic attraction force. Comb-drives are a more advanced type (Tang et al., 1989), where two frames with interdigitated fingers slide within each other through electrostatic forces, as shown in Figure 6.5. Comb-drives are widely used both for actuation but also for capacitive sensors (Krijnen et al., 2003); in grippers, they can play either role. The comb-drive has several advantages over the simple plate capacitor; they allow a larger stroke and larger force, since they do not suffer from snap-in instability in the same way as plate capacitors. They can exhibit side instability if the gaps are too small or the overlap of the pins is too large (Krijnen et al., 2003). Depending on the actuator design, comb-drives have a nearly constant force, whereas plate capacitor actuators are nonlinear.

6.3.1 PLATE CAPACITOR DRIVE

In the following, we will discuss how the movement—the actuation—depends on an electrical voltage, as most of electrostatic actuators used for manipulation are voltage

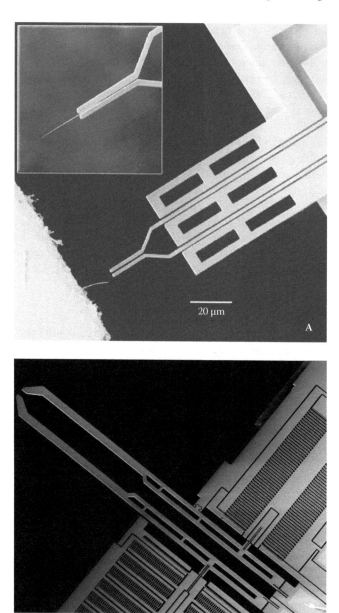

FIGURE 6.5 Two examples of electrostatic actuators: (A) a plate-capacitor-driven micro-gripper with both open and close capability and (B) a more advanced gripper with lateral comb-drives and integrated capacitive position readout (Courtesy of Femtotools).

controlled. Consider two capacitor plates with a gap d and area $A = hw$, where h and w are the height and width, respectively. The capacitance is $C = \varepsilon A/g$, where ε is the permittivity (Figure 6.6), which is very close to the dielectric constant $\varepsilon \approx \varepsilon_0$ in air.

In the following, we obtain the relationship between gap size g, force F, and bias voltage V applied to the actuator, outlining the derivation given by Senturia (2000). The energy stored in an electrostatic capacitor is given by

$$W(Q,g) = \frac{Q^2 g}{2\varepsilon A} \tag{6.3}$$

Thus, the energy can be changed by varying the charge Q, the gap, or the area. By adding a spring with spring constant k to one of the capacitor plates, we now have an actuator. This is illustrated schematically as a spring k holding the upper plate in Figure 6.7a.

For a *voltage*-controlled actuator, the electrical *co-energy* rather than the energy of the capacitor is considered, which is defined as (Senturia, 2000):

$$W^*(V,g) = QV - W(Q,g), \tag{6.4}$$

In differential form, this can be rewritten as:

$$dW^*(V,g) = QdV - Fdg, \tag{6.5}$$

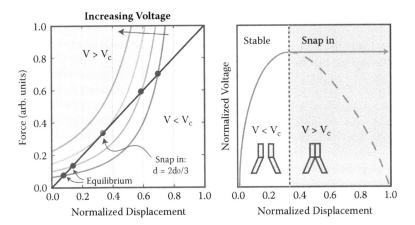

FIGURE 6.6 **(See color insert.)** The mechanical (black curve) and electrostatic (colored) forces are plotted for a voltage-controlled electrostatic actuator. For low voltages, the curves cross at two points, where the lower is the mechanical equilibrium. As the voltage increases, the stable and unstable solutions merge into one at the "snap-in" voltage (blue curve), beyond which there are no stable solutions; that Is, the structure collapses. By plotting the voltage against the normalized displacement, the snap-in occurs at $dV/dg = 0$, or at 2/3 of the initial gap size g_0.

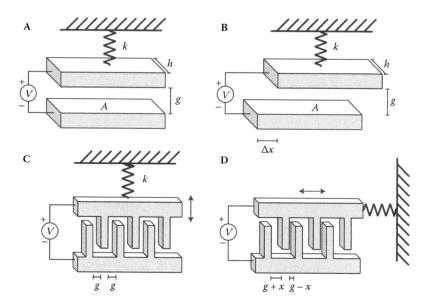

FIGURE 6.7 (A) Illustration of an electrostatic actuator based on a plate capacitor with one plate suspended in a spring with spring constant k. (B) The same actuator where the plates are sliding, rather than moving away/toward each other. (C) A comb drive with longitudinal actuation, that is, the areal overlap changes. (D) A comb-drive with transversal actuation; here, half the gaps increase while the others decrease in size. (Adapted from Sahu, B., Taylor, C. R., and Leang, K. K. *Journal of Manufacturing Science and Engineering—Transactions of the ASME*, 132., 030917-4, 2010. With permission.)

showing that both electrostatic and mechanical energy can be stored in the actuator. From this, we see that the charge Q and the force F can be found from:

$$Q = -\frac{\partial W^*(V,g)}{\partial V}\bigg|_g$$

$$F = -\frac{\partial W^*(V,g)}{\partial g}\bigg|_V \tag{6.6}$$

Using the first expression (fixed gap), we can calculate the co-energy

$$W^*(V,g) = \int_0^V QdV = \frac{\varepsilon A V^2}{2g}, \tag{6.7}$$

where $Q = CV = \varepsilon A V / g$. The force can then be written as

$$F = -\frac{\partial W^*(V,g)}{\partial g}\bigg|_V = \frac{\varepsilon A V^2}{2g^2}. \tag{6.8}$$

In the actuator, equilibrium between the electrostatic force F_{el} and the mechanical force F_{mech} defines the actual gap d between the plates. The unbiased gap is g_0 and the mechanical force given by Hooke's law is $F_m = -kz = -k(g - g_0)$, with k being the spring constant of the actuator. From $F_m = F_e$, we get for a displacement in the direction of the gap:

$$g = g_0 - \frac{\varepsilon A V^2}{2kg^2},$$

As the voltage increases, not only does the gap decrease, but the rate of change increases. This ultimately leads to the earlier mentioned situation called pull-in or snap-in instability. This can be solved graphically, by comparing the electrostatic force F_m to F_e as a function of the normalized displacement $(g - g_0) / g_0$ for different voltages V as shown in Figure 6.6. For small voltages, there are two solutions, of which the solution with the smallest displacement z is the stable equilibrium point. As seen in the Figure 6.6a, there is a certain critical voltage V_c, where only one, unstable solution exist. Beyond this voltage, the mechanical restoration force cannot hold back the electrostatic force, and the actuator pulls in, which occurs at the snap-in voltage (Senturia, 2000):

$$V_{snap-in} = \sqrt{\frac{8}{27} \frac{kg_0^3}{\varepsilon A}}$$

One way to avoid the pull-in instability is to operate the actuator in a different way: by letting the capacitor plates slide rather than move against each other, which changes the capacitance as well. If a lateral offset Δx is introduced, corresponding to the plates sliding with respect to each other at a constant gap g_0, the transversal force turns out to be independent of Δx as well as of the area:

$$F_{eltran} = -V^2 \frac{\varepsilon h}{2g_0},$$

The plate capacitor formula is not very accurate unless the gap is small compared to the dimensions (width and height) in the lateral direction, since here the fringe field must be taken into account (Senturia, 2000). Mølhave and coauthors (2006) demonstrated pick and place of silicon nanowires using a parallel plate gripper. In this work, it was, however, clear that the force delivered by this type of gripper is insufficient for a range of manipulation tasks; for a larger actuation range and gripping force, electrostatic comb-drive actuators and electrothermal actuators are preferable.

6.3.2 COMB-DRIVES

Comb-drives are by far the most widely used type of electrostatic actuator for grippers and tweezers, as well as generally in MEMS devices. Figure 6.5b shows a SEM image of a comb-driven microgripper, while Figure 6.7c shows a schematic of a comb-drive. While the force per area is smaller than for a plate-capacitor drive (Johnstone and Paramswaran, 2004), it is easier to design comb-drives with a reasonable compromise between actuation range and force. The comb-drive comes in two variants, depending on how the pins move with respect to each other. In the longitudinal comb-drive, the fingers slide into each other, causing the overlap area A of each pair of pins to change (Figure 6.7c). If the fingers have a height h and an overlap x with other fingers, the area is $A = xh$. Each of the N fingers form two plate capacitors with the fingers of the opposing comb (Krijnen et al., 2003), so the actuation can be written as

$$\Delta x = N \frac{\varepsilon h V^2}{kg^2},$$

and hence, the total force between the two capacitor structures of a comb-drive with N fingers becomes independent of x.

This is very convenient since the electrostatic force is then constant over large travel ranges (Krijnen et al., 2003). Increasing the force is straightforward: the number of gaps and pins is increased, which in turn makes the actuator more bulky. The design of the suspension of the spring is also very important in determining the actuation range, stability, and available force of the actuator (Legtenberg et al., 1996).

The transversal forces in the longitudinal drive cancel each other out, but the equilibrium position is unstable; this can lead to instabilities, and ultimately failure, if the gap between the actuator is made too small compared to the rigidity of the structures.

This lateral attractive force is exploited in the transversal drive; see Figure 6.7. Here, the distance between the pins changes; this type of drive has some of the same drawbacks of nonlinearity as plate capacitor-like drives (Sun et al., 2005), as well as the pull-in instability.

6.3.3 ELECTROSTATIC GRIPPERS FOR BIOMANIPULATION

Electrostatic grippers must be designed to avoid snap-in, to avoid current passing through the sample and to avoid high electrical fields near the tips, which might generate unwanted liquid motion due to electrophoresis, dielectrophoresis, or electrolysis (Desai, 2007). As seen in Figure 6.6b, the displacement of the actuator is parabolic for small bias voltages, so not being allowed to use large voltages can be serious, as this can significantly reduce the available actuation range.

One option is to generate the deflection in a remote comb drive actuator and transfer the force to the gripper arms as shown in the device in Figure 6.5, thus separating the end-effector region from the actuator region. Others focus on carefully designing the actuator to provide a sufficient force (1 nN – 10 μN) for manipulation of biological specimens, small operating voltages (< 5–10 V) and large enough actuation, even when immersed. Mukundan and Pruitt (2009) used high-frequency voltages to operate an electrostatic actuator in an ionic solution, largely eliminating electrolysis as well as electric shielding by the ions, similar to ac dielectrophoresis (Dimaki and Boggild, 2004). The rapid shifting of the voltage polarity make the ions in solution oscillate back and forth, reducing the amount reaching the electrodes strongly. The authors (Mukundan and Pruitt, 2009) studied the mechanical properties of living cells and their response to mechanical stimuli generated by the actuators.

The Swiss company Femtotools (Femtotools) produces a range of electrostatic tweezers and has demonstrated their use in vacuum, air, as well as a liquid environment; see Figure 6.5b. The grippers have built-in force sensing with a resolution down to 0.0006 µN. Neild and co-workers (2006) used this type of gripper to first arrange microparticles in a line inside a narrow fluidic microchannel by ultrasonic excitation. The acoustic waves trapped the particles in the center of the channel, which is open to the surrounding air in one end. The gripper then penetrated the air–water interface to pick up the 74 µm particles one by one. The authors found that the ultrasonic agitation of the surrounding water also led to release of a stuck particle, similar to the findings of Haliyo et al (2002).

Nanotube Nanotweezers by Kim and Lieber (1999), are perhaps the most spectacular nanomanipulation tool in terms of the gripper size and concept. Two carbon nanotubes were attached to the sides of a glass capillary by manipulation and glue. A careful deposition of metal on both sides of the capillary tube then connects the carbon nanotubes electrically, forming a pair of electrostatic nanotweezers; when applying a bias voltage, the nanotubes are attracted to each other. A variant of this type of probe was used by Watanabe and co-workers (Watanabe et al., 2001; Shimotani et al., 2003) to pick up individual DNA molecules and even perform electrical three-terminal measurements on these, while being held by the carbon nanotubes. Although nanowires, nanoparticles, and DNA molecules have been manipulated using this type of tweezers, the force that can be applied by the thin, flexible carbon nanotube is limited. Also, fabrication of such tweezers is performed manually under an optical microscope, and is presumably highly dependent on the skills of the operator.

6.4 ELECTROTHERMAL GRIPPERS

Electrothermal actuation occurs when two objects with either different temperature (electrothermal actuator) or thermal expansion coefficient (bimorph) are used to create a motion as a consequence of a change in temperature or a certain temperature distribution. One of the early types of electrothermal MEMS actuators is the Guckel actuator (Johnstone and Paramswaran, 2004), also called the *thin-beam thick beam* actuator (see Figure 6.8a) (Pelesko, 2002). This is a U-shaped cantilever where the two arms have different cross sections. As an electrical current is passed through this asymmetric U-loop, the larger current density in the thin beam will make it hotter than the wide beam, and therefore make it elongate more, due to thermal expansion. Although the relative expansion is only on the order of 10×10^{-6} per Kelvin, an appropriate design can amplify the shape change into an actuation of a useful magnitude. In the Guckel actuator, the mechanical bridge at the end of the gripper connects the two beams and forces the structure to be twisted sideways. This principle has been used for a variety of different grippers (Yan et al., 2003; Chronis and Lee, 2005; Sardan et al., 2008b), where the three-beam actuator proposed by Mølhave and Hansen (2005) is an interesting variant (Figure 6.8b). In this gripper, the current can be passed through the two outer beams (see Figure 6.8), which means that these will heat up compared to the inner beam, and force the gripper to twist toward the right. Alternatively, by passing the current in through

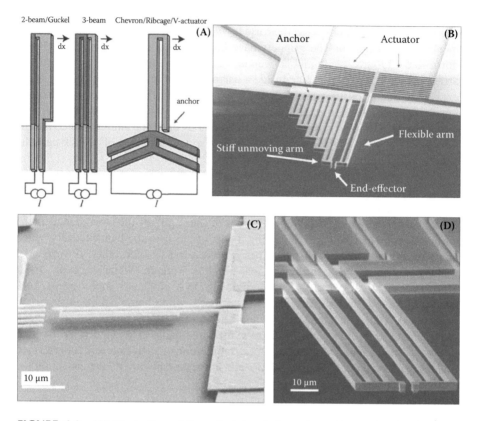

FIGURE 6.8 (A) Illustration of the principle of electrothermal actuation implemented in three different geometries: two-beam (Guckel), three-beam, and ribcage/chevron. The regions primarily heated up by the electrical current are shown in dark gray. SEM images of (B) a Chevron actuator (Reprinted from Carlson, et al. *Nanotechnology*, 18, 2007. IOP Publisher. With permission) (C) a Guckel actuator (Reprinted from Lee, et al. *Journal of Micromechanics and Microengineering*, 15, 322–327, 2005. Copyright 2005, IOP Publishing. With permission) and (D) a three-beam actuator (Andersen, et al. *IEEE Transactions on Nanotechnology*, 8, 76–85, 2009) are shown.

the two inner beams instead, the arm bends the other way, allowing the gripper to both open and close. This type of gripper can also be used as an electrostatic gripper, by applying a voltage difference between each of the two arms (Andersen et al., 2009).

One issue common to most Guckel-type grippers is that the expansion is generated in the gripper arm itself, which means that the highest temperature will be near the end-effectors and potentially damage the object to be manipulated. By modifying the shape of the thin-beam thick-beam actuator, Li and Uttamchandani (2004) managed to flatten out the temperature distribution and thereby achieve smaller maximal temperatures while maintaining the performance. Park et al. (2010) demonstrated a submicron metallic Chevron gripper, capable of manipulating submicron structures (Figure 6.8c).

Another very successful design is the ribcage, or the Chevron actuator; see Figure 6.8b. Here, two beams are opposing each other at an angle; when they expand, they will push against each other, forcing them both to move in the direction of the tilt. The V-shaped structure amplifies the actuation significantly.

Similar to electrostatic comb-drives, Chevron or V-shaped actuators can easily be scaled up to provide a higher force, and positioned far away from the end-effectors to affect the manipulated object less, at the cost of taking up more space on the chip. The US company Zyvex offers a wide range of electrothermal grippers with an actuation range of up to 50 µm, in addition to advanced multiprobe 3D manipulator systems for controlling electrical probes and grippers inside a scanning electron microscope. The grippers are based on the work by Skidmore and coworkers (Lee et al., 2003; Oh et al., 2003).

6.4.1 THE U-LOOP AND THE CHEVRON DRIVE

In this section, we take a look at the physics behind the electrothermal actuator. The discussion here is rather rudimentary; a more comprehensive analysis is given in work by Huang and Lee (1999) and Hickey et al. (2002). The electrothermal gripper uses electrical current to generate heat, which in turn leads to expansion of the material, and which is finally used to change the shape in a way that results in a force acting on the object. A very simple case is a U-shaped beam, which resembles several types of actuators shown in Figure 6.8: the two-beam actuator (Guckel actuator), the three-beam actuator, as well as the Chevron actuator, if the arms of the U-actuator are pulled into a V-shape. Each arm has the length L, and a homogeneous cross section A; see Figure 6.9. In the following, we consider the contributions to the temperature change of a small section of the loop, calculate the temperature distribution $T(x)$ for the full loop, and from that the total elongation dL, which we use to estimate the actuation achievable with the Chevron actuator (V-shaped).

Ohm's law, $\Delta V = RI$, relates the current I generated by a voltage difference ΔV across a conductor with resistance R, which generates an electrical charge flow $I = dq/dt$. The resistance R can be expressed as $R = \rho L/A = L/\sigma A$, for a conductor with cross-sectional area A, length L, and electrical resistivity $\rho = \sigma^{-1}$; here σ is the electrical conductivity. We can now write Ohm's law as:

$$\frac{dq}{dt} = \sigma \frac{A}{L} \Delta V$$

Similarly, the heat flow dQ/dt generated by a temperature difference ΔT across a heat conductor with area A and length L, and heat conductivity κ is given by

$$\frac{dQ}{dt} = -\kappa \frac{A}{L} \Delta T$$

Here, the negative sign indicates that the heat flows toward the region with the lowest temperature. If we divide the U-shaped beam into small elements of length dx, and

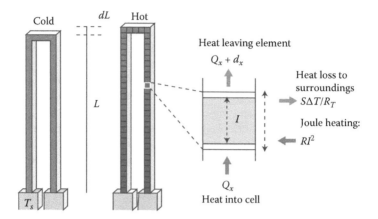

FIGURE 6.9 **(See color insert.)** A U-shaped beam of length L, connected to two room temperature contacts of temperature T_s. The heat flow in and out of the infinitesimal element dx, Q_x and Q_{x+dx}, as well as the heat generated by current heating, RI^2 and the heat exchange with the surroundings, $S\Delta T / R_T$ (see text).

no current is flowing, the heat flow into the element at x and out of the element at $x + dx$ is the same; see Figure 6.9. Substituting $L = dx$, we get

$$-\kappa A \frac{dT}{dx}\bigg|_x = -\kappa A \frac{dT}{dx}\bigg|_{x+dx}$$

Let us now consider both the contribution by an electrical current, which transfers heat to the element, and the surroundings, to which the element loses heat. If a current I is passed through the element, an electrical power RI^2 will be dissipated in the section. The generated heat can be written in terms of the current density $J = I/A$ and the electrical resistivity $\rho = RA / dx$:

$$RI^2 = \rho \frac{dx}{A}(AJ)^2 = \rho AJ^2 dx$$

The heat exchange with the surroundings for an element with width w and length dx can be described by the term (Yan et al., 2003) $Sw(T - T_s)dx/R_T$, where S is a form factor that depends on the shape and area of the actuator; T_s is the surroundings temperature; and R_T is the combined thermal resistance between the actuator structure to the substrate, and to the air or liquid environment in touch with the actuator. By dividing with the cross-sectional area $A = wt$ on both sides, the full heat budget for the section can then be written:

$$\frac{dT}{dx}\bigg|_{x+dx} - \frac{dT}{dx}\bigg|_x = \frac{S(T - T_s)dx}{\kappa R_T t} - \frac{\rho J^2}{\kappa} dx$$

For manipulation in vacuum or air, the heat conduction to the surroundings is sometimes neglected (Huang and Lee, 1999), just as radiation loss gives a relatively small contribution even at a temperature of 1000°C (Hickey et al., 2002). This does not hold for liquid, as the thermal resistance to liquid is much smaller than to air, leading to a significant exchange of heat with the surroundings, that is, R_T being small. For the purpose of simplicity, we however consider an electrothermal actuator in vacuum. Rearrangement of the terms in Equation 6.16 in the limit $dx \to 0$ gives the heat conduction continuity equation (Molhave and Hansen, 2005; Pelesko, 2002):

$$\frac{d^2T}{dx^2} = -\frac{\rho J^2}{\kappa}$$

This is easily solvable for a U-loop with total length L (see Figure 6.9), so that each of the arms has length $L/2$ (the bridge in the U-loop is neglected here), and using the boundary conditions that the temperature at the two bases of the U-loop is at room temperature, $T(0) = T(L) = T_s$. The solution is then a parabolic temperature distribution:

$$T(x) = \frac{\rho J^2}{2\kappa}(xL - x^2) + T_s$$

Using $RA = \rho L$ and $J = I/A = V/AR$, this can be rewritten in terms of only known parameters:

$$T(x) = \frac{V^2}{2\rho\kappa L^2}(xL - x^2) + T_s$$

The temperature must be highest at the center of the U-loop, that is, $T_{max} = T(L/2) = V^2 / 8\rho\kappa + T_s$. For a thin-beam thick beam actuator with a flexure point, the hottest point is shifted toward the thin arm, as shown in Figure 6.10.

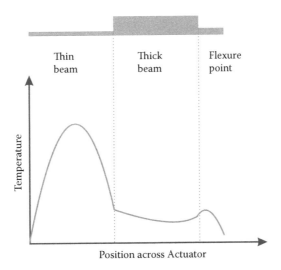

FIGURE 6.10 Schematic of the temperature distribution calculated as a function of position along the actuator, which is shown unfolded.

If the temperature in a beam is increased *homogeneously*, $\Delta T = T - T_0$, it will elongate by $\Delta L = L\alpha\Delta T$, where L is the initial length. The thermal expansion coefficient $\alpha = (dL / dT) / L$ can be regarded as temperature independent for many materials, such as for instance metals. The thermal expansion coefficient for metals on the order of 10^{-5} K^{-1}, meaning that a 100 K temperature change will only elongate the structure by 0.1%. Using a design such as the Guckel or V-shaped actuator, this small elongation can be amplified, at the cost of a reduced force. As a rule, the larger the actuation range, the smaller the applicable force.

For each of the volume elements shown in Figure 6.9, the change of size as the temperature is increased from T_s to T_h given by the thermal expansion coefficient α of the material:

$$L(T_h) = L(1 + \alpha(T_h - T_s))$$

As we just found, Equation 6.18, the temperature is not constant but rather a distribution $T(x)$, since the beams are connected to large contacts at the base, where the temperature is assumed to be constant T_s. For the U-shaped beam of total length L, we integrate the thermal expansion along each arm:

$$L_i = L + \int_0^L \alpha(T(x) - T_s)dx = L + \frac{V^2}{4\rho\kappa L^2}\int_0^L (xL - \tfrac{1}{2}x^2)dx$$

to get the elongation of the U-beam:

$$L(V) = \frac{L}{2}\left(1 + \frac{\alpha\rho J^2}{12\kappa}L^2\right)$$

By again using $J^2 = V^2 / R^2A^2$ and $R^2 = \rho^2L^2 / A^2$, we get the solutions for the elongation, keeping either the voltage or the current constant:

$$L(V) = \frac{L}{2}\left(1 + \frac{\alpha}{12\kappa\rho}V^2\right), \text{ and } L(I) = \frac{L}{2}\left(1 + \frac{\alpha\rho L^2}{12\kappa A^2}I^2\right)$$

Hence, the elongation of the U-shaped beam is inversely proportional to the cross-sectional area *squared* in constant-current mode. The maximal voltage $V_{max} = (\Delta T_{max} 8\kappa\rho)^{1/2}$ is given by Equation 6.19, which then gives the elongation at the maximal temperature:

$$\Delta L = \frac{L_{max} - L}{2} = \frac{L}{2}\left(\frac{\alpha 8\Delta T_{max}\kappa\rho}{12\kappa\rho}\right) = \frac{\alpha L\Delta T_{max}}{3}$$

If we assume the beam to be $L = 200$ μm in total (left plus right arm), we get by insertion a 0.34 μm elongation at a temperature near the melting temperature of silicon. If we, however, only accept a more biocompatible temperature increase of 10°C, the elongation is just 9 nm. Taking heat conductance to the surrounding into account will reduce the effect even more, as this will lower the temperature of the structure.

In the Chevron actuator, the U-loop is stretched out to a V-shape, as shown in Figure 6.11. Because the "mirror" leg of the V prevents the actuator from stretching in the x-direction, it is pushed upward in the y-direction. One can get an idea of the geometrical amplification effect by considering Figure 6.11, where trigonometry without any concern for the mechanical properties gives: $0 = dy^2 + 2ydy - dL^2 - 2LdL$. Due to the very small elongation, $dL \ll L$, we can approximate this by $dy2 \to 0$; $dL^2 \to 0$ whereby the actuator is moving $dy = dL(L/y)$ in the y direction. So, as the V becomes more and more outstretched, y is reduced, which increases the amplification factor L/y. We can thus write a crude expression for the actuation of a Chevron actuator:

$$\Delta y = \frac{L}{y} \left(\frac{\alpha L \Delta T_{max}}{3} \right) = \frac{1}{24} \frac{L^2}{y} \frac{\alpha V^2}{\kappa \rho}$$

This estimate altogether neglects elasticity theory; for instance, the case of opposing beams ($y = 0$) should give infinite actuation; in practice, the beams would just buckle. The actual mechanical actuation response to a given temperature distribution is calculated using continuum mechanics by Smith et al. (2001), who compares models and measurements of electrothermal V-actuators and electrostatic comb-drives, as well as in many MEMS papers (Molhave and Hansen, 2005; Yan et al., 2003; Senturia, 2000; Landau, 1960). Calculation of the current distribution, temperature distribution, thermal expansion and, finally, the mechanical response for a given actuator structure is often only possible with finite element methods (Yan et al., 2003; Luo et al., 2005; Neild et al., 2006; Sardan et al., 2008b). One complication for silicon is that both the electrical resistivity as well as the thermal conductance depend strongly on the doping level and temperature (Sardan et al., 2008a).

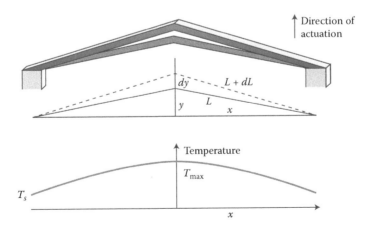

FIGURE 6.11 The "Chevron" actuator expands in the y-direction when the beams elongate in the x-direction. The temperature distribution is parabolic, with the maximal temperature at the apex.

6.4.2 TOPOLOGY-OPTIMIZED ACTUATORS

A far more radical approach to improving the design is topology optimization, developed by Bendsøe and Sigmund (2002). This is a general method for optimization of a shape of a device or an object to best meet the requirements of a certain application, covering a large number of different physical and engineering fields. In short, the system calculates what the shape should be in order to fulfill certain outer performance goals and constraints as best as possible. This can be thought of as an iterative finite element simulation, where a clever optimization method tunes the shape of the structure slightly between each successive iteration. The constraints would for electrothermal grippers be the number and placement of electrodes, and the amount of material available for the algorithm to redistribute; see Figure 6.12. The equations guiding the optimization for the electrothermal actuator are partial differential equations related to the electrical, thermal, and mechanical domains. Most notably, the outcomes of this type of optimization are very different from the typical human-made designs, sometimes also with vastly better performance. Jonsmann et al. (1999) reported the first use of topology optimization for electrothermal actuators, and Sardan and co-workers (2008a; 2008b) found that for similar overall size and use conditions, improvements by a factor of 10–100 in terms of actuation force could be achieved while maintaining the same actuation range as for a three-beam gripper. This allowed the gripper to be miniaturized by a factor of 5–10 while still providing a reasonable force (Cagliani et al., 2010). The gripper was used for automated manipulation of carbon nanotubes inside a scanning electron microscope of structures down to tens of nanometers (Sardan et al., 2008a; Cagliani et al., 2010).

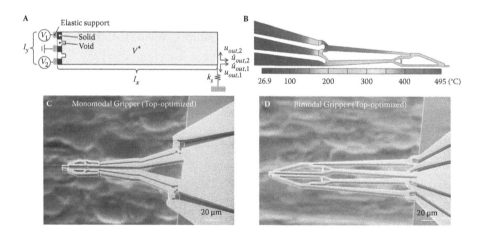

FIGURE 6.12 (See color insert) Topology optimization. The upper panel shows the design domain and the constraints: the overall size, the available amount of material, and the position of electrodes. The goal is to maximize simultaneously the spring constant and the actuation range ($u_{out,1}, u_{out,2}$) of the lower-right corner of the structure. The upper-right corner shows finite element calculations of the temperature distribution using COMSOL. The lower SEM images show a monomodal gripper (can close) and a bimodal gripper (can close as well as open). (Courtesy of Özlem Sardan.)

Topology optimization has been used for a range of other micromanipulation tools (Helal et al., 2009, 2010; Rubio et al., 2009). This method could be very useful for adapting electrothermal grippers to biological applications, since larger force and actuation at the same temperature will greatly improve the usefulness of such grippers in a temperature-sensitive environment.

6.4.3 ELECTROTHERMAL GRIPPERS FOR BIOMANIPULATION

As mentioned before, one serious problem for electrothermal grippers in saline solutions is electrolysis, as well as heating of the liquid. Kim et al. (2008) showed an electrothermal Chevron-type silicon gripper with capacitive two-axis force feedback, capable of resolving gripping forces down to 20 nN and contact forces down to 40 nN. In this case, only the gripping arms were immersed in the liquid, leaving the hotter actuator region outside. The authors reported a strong increase in the effectiveness of the manipulation process due to the force sensing capability; the contact force sensor gave a fast and precise indication of when the gripper reaches the substrate within the droplet, and the gripper force allowed safe handling of living cells. A 100 nN gripping force resulted in a 15% deformation of porcine aortic valve interstitial cells (PAVICs); however, the different size and stiffness of the cells required a different bias voltage to safely manipulate different cells. A closed-loop control was therefore used to maintain a constant gripping force. The temperature rise was kept at a level where the cells could survive.

To counteract the heating and electrolysis problems, several groups have successfully fabricated hybrid grippers, consisting of a metal and a polymeric part. SU-8 is a suitable choice to provide the thermal expansion due to its relatively good mechanical properties, with a Young's modulus of 5 GPa and a high thermal expansion rate, which is about $52 \times 10^{-6} K^{-1}$ compared to $3 \times 10^{-6} K^{-1}$ for silicon and ($12 \times 10^{-6} K^{-1}$) for Au. Chronis and Lee (Chronis and Lee, 2005) demonstrated a thermally actuated SU-8 polymer Guckel-type gripper, which could grip and position a 10 μm cell in a saline solution. The gripper could open 11 μm with a temperature change less than 32 K at an applied voltage of 2 V, very near the electrolysis threshold (Neagu et al., 1998). Colinjivadi et al. (2008) reported viable cell handling using a metallic heater integrated on top of a SU-8 gripper structure, where the heat was transferred to a Chevron-type actuator. Careful finite element simulation allowed the authors to tune the operation window to below 1 V of bias voltage, a temperature increase estimated to be less than a few degrees centigrade, and an actuation range of tens of microns.

Finally, Elbuken and co-workers (2008a; 2008b) showed a working SU-8 gripper intended for single cell handling, which was heated by focusing a laser spot on the actuator region. The advantage is that the stacked layer structure is avoided, with metal heaters integrated with the SU-8 polymeric gripper structure; the adhesion between metal and SU-8 is poor. Moreover, the bilayer structure makes deformation of stress more likely due to the different expansion coefficients, which can lead to out-of-plane bending. It remains to be shown how well the concept works in a liquid environment, where the laser might give rise to unwanted heating.

Hashiguchi et al. (2003) used an electrothermal nanogripper to extract DNA from aqueous solution and attempted electrical measurements without being able to detect any conductance. The gripper was in this case was not used to apply mechanical force, but simply through dielectrophoresis and surface tension effects to arrange DNA at one or both nanogripper tips.

6.5   PIEZOELECTRIC AND OTHER TYPES OF ACTUATION

Whereas piezoelectric actuators have been used widely for micropositioning, integrated two-finger designs (tweezers, grippers) are more complex to realize. Two tips operated by individual multiaxis actuators were used to move human white blood cells (Tanikawa and Arai, 1999), but there have also been successful attempts to integrate a piezoelectric-actuated gripper in a chip; Chen and co-workers reported a piezoelectric gripper with built-in piezoresistive force sensor on the sidewall of the gripper arm (Chen et al., 2009). While issues such as reliability, creep, and stability (Sahu et al., 2010) have to some extent prevented piezoelectric grippers from use except for demonstrations, improvements of fabrication and integration technology could make this type of gripper viable for biomanipulation. Other actuator types that might become relevant for biomanipulation are, for instance, pneumatic end-effectors (Lu and Kim, 2006) and ionic conducting polymer films (Zhou et al., 2003).

6.6   SUMMARY: FEASIBILITY OF GRIPPER BIOMANIPULATION

In this chapter, we have reviewed general aspects of MEMS-based microgrippers and the state of the art of gripper-based biomanipulation, with the aim of at least partly answering the question: are grippers of any use for microbiology? The capillary tube is extremely simple and versatile, and can inject, extract, and move cellular structures. Optical tweezers move single cells without physical contact, allowing delicate mechanical operations and studies to be carried out. What does the gripper have to offer in comparison?

The microgripper can be seen as a clumsy, brute-force approach that has none of the elegance of holding a cell with a beam of light, or moving biological objects solely with electrical or magnetic fields. While gripper biomanipulation is still in its infancy, some remarkable experiments and demonstrations have been performed, which show some promise. We have here focused on some of the more advanced techniques for physical 3D manipulation of biological objects, which here applies to micron-size or larger objects; manipulation of subcellular objects with grippers is extremely difficult; the successful extraction of DNA from aqueous solution employing adhesive rather than mechanical forces to "grab" DNA. It is doubtful that direct manipulation of biological objects through opposing forces—grippers—is ever going to be important on the submicron scale.

Thus, manipulation of cells and perhaps bacteria appears to be the immediate application area of interest. The most promising types of structures are electrothermal grippers where the expanding part is made of the polymer SU-8, as this seems to be very effective in providing low operation voltages and heat load, which is critically important to prevent damage to living cells. Likewise, electrostatic actuators operating at high frequency seem to be effective in preventing electrolysis and other harmful effects. Simply keeping the gripper outside is also an option; inserting the gripper end-effectors through the air–liquid boundary is an interesting alternative, as electrostatic actuators by all means work better in air or vacuum than in saline solution. There are several routes to accurate position detection providing force feedback, and strongly improving the chances of not destroying the cells; here, capacitive readout is perhaps the most convenient, well developed, and precise, as it promotes force measurement in the nanonewton regime.

Apart from creating the tools, development of control systems is another challenge. Automated manipulation in solution is already difficult; the added complexity of picking and placing will demand further technological development. Possibly the integration of force feedback in combination with haptic user interfaces could make handling safer and easier. Issues such as teleoperation, haptic feedback, and even virtual reality control will become more relevant, as reliable 3D manipulation becomes available on a regular basis. Finally, automation will be necessary to some degree; if not full-fledged automated manipulation, then some ability to define and play back action sequences to eventually create a user environment where nonexperts can carry out manipulation tasks on a routine basis. In such a scenario, researchers and engineers working with biological samples could take advantage of a significant expansion of the toolbox, as the shape, size, function, and materials of grippers and tweezers, as this chapter has hopefully demonstrated, is highly customizable, and it should be possible to adapt them to a wide range of biological applications.

REFERENCES

Ahmad, M. R., Nakajima, M., Kojima, S., Homma, M., and Fukuda, T. (2008). In situ single cell mechanics characterization of yeast cells using nanoneedles inside environmental SEM. *IEEE Transactions on Nanotechnology,* 7, 607–616.

Andersen, K. N., Petersen, D. H., Carlson, K., Molhave, K., Sardan, O., Horsewell, A., Eichhorn, V., Fatikow, S., and Boggild, P. (2009). Multimodal electrothermal silicon microgrippers for nanotube manipulation. *IEEE Transactions on Nanotechnology,* 8, 76–85.

Bendsøe, M. P. and Sigmund, O. (2002). *Topology Optimization.* Berlin: Springer.

Bhushan, B. (2010). *Springer Handbook of Nanotechnology.* New York: Springer.

Boggild, P., Hansen, T. M., Tanasa, C., and Grey, F. (2001). Fabrication and actuation of customized nanotweezers with a 25 nm gap. *Nanotechnology,* 12, 331–335.

Cagliani, A., Wierzbicki, R., Occhipinti, L., Petersen, D. H., Dyvelkov, K. N., Sukas, O. S., Herstrom, B. G., Booth, T., and Boggild, P. (2010). Manipulation and in situ transmission electron microscope characterization of sub-100 nm nanostructures using a microfabricated nanogripper. *Journal of Micromechanics and Microengineering,* 20. 035009.

Carlson, K., Andersen, K. N., Eichhorn, V., Petersen, D. H., Molhave, K., Bu, I. Y. Y., Teo, K. B. K., Milne, W. I., Fatikow, S., and Boggild, P. (2007). A carbon nanofibre scanning probe assembled using an electrothermal microgripper. *Nanotechnology,* 18. 345501.

Chen, T., Chen, L. G., Sun, L. N., and IEEE (2009). Piezoelectrically Driven Silicon Microgrippers Integrated with Sidewall Piezoresistive Sensor. *ICRA: 2009 IEEE International Conference on Robotics and Automation,* Vols. 1–7.

Chronis, N. and Lee, L. P. (2005). Electrothermally activated SU-8 Microgripper for single cell manipulation in solution. *Journal of Microelectromechanical Systems,* 14, 857–863.

Colinjivadi, K. S., Lee, J. B., and Draper, R. (2008). Viable cell handling with high aspect ratio polymer chopstick gripper mounted on a nano precision manipulator. *Microsystem Technologies: Micro- and Nanosystems-Information Storage and Processing Systems,* 14, 1627–1633.

Desai, J. P., Pillarisetti, A., and Brooks, A. D. (2007). Engineering approaches to biomanipulation. *Annual Review of Biomedical Engineering*, 9, 35–53.

Dimaki, M. and Boggild, P. (2004). Dielectrophoresis of carbon nanotubes using microelectrodes: A numerical study. *Nanotechnology*, 15, 1095–1102.

Duch, M., Lopez, M. J., Gomez, R., Esteve, J., and Plaza, J. A. (2009). *Microneedles Electrodes for Living Cells*. Proceedings of the 2009 Spanish Conference on electron devises. Santiagode compostela, Spain 2009.

Elbuken, C., Gui, L., Ren, C. L., Yavuz, M., Khamesee, M. B., and IEEE (2008a). *A Monolithic Polymeric Microgripper with Photo-Thermal Actuation for Biomanipulation*.

Elbuken, C., Gui, L., Ren, C. L., Yavuz, M., and Kharnesee, M. B. (2008b). Design and analysis of a polymeric photo-thermal microactuator. *Sensors and Actuators A-Physical*, 147, 292–299.

Fatikow, S., Eichhorn, V., Stolle, C., Sievers, T., and Jahnisch, M. (2008). Development and control of a versatile nanohandling robot cell. *Mechatronics*, 18, 370–380.

Fatikow, S. and Rembold, U. (2002). *Microsystem Technology and Microrobotics (Microsystem Technology and Microrobotics)*. Heidelberg, Germany: Springer.

Femtotools. www.femtotools.com.

Fernandez-Varea, J. M. and Garcia-Molina, R. (2000). Hamaker constants of systems involving water obtained from a dielectric function that fulfills the f sum rule. *Journal of Colloid and Interface Science*, 231, 394–397.

Hashiguchi, G., Goda, T., Hosogi, M., Hirano, K., Kaji, N., Baba, Y., Kakushima, K.., and Fujita, H. (2003). DNA manipulation and retrieval from an aqueous solution with micromachined nanotweezers. *Analytical Chemistry*, 75, 4347–4350.

Helal, M., Sun, L. N., and Chen, L. G. (2010). Three-dimensional optimal material distributions for micro-gripper with straight-line path and parallel movement arms. *Materials Testing—Materials and Components Technology and Application*, 52, 174–181.

Helal, M., Sun, L. N., and Chen, L. G. (2009). Optimal material distribution of a microgripper manipulator with straight-line path and parallel movements. *Materials Testing—Materials and Components Technology and Application*, 51, 794–801.

Hickey, R., Kujath, M., and Hubbard, T. (2002). Heat transfer analysis and optimization of two-beam microelectromechanical thermal actuators. *Journal of Vacuum Science and Technology A—Vacuum Surfaces and Films*, 20, 971–974.

Huang, H. B., Sun, D., Mills, J. K., Cheng, S. H., and IEEE (2008). Integrated vision and force control in suspended cell injection system: Towards automatic batch biomanipulation. *2008 IEEE International Conference on Robotics and Automation, Vols 1–9*.

Huang, Q. A. and Lee, N. K. S. (1999). Analytical modeling and optimization for a laterally-driven polysilicon thermal actuator. *Microsystem Technologies—Micro- and Nanosystems Information Storage and Processing Systems*, 5, 133–137.

Hubbard, N. B., Culpepper, M. L., and Howell, L. L. (2006). Actuators for micropositioners and nanopositioners. *Applied Mechanics Reviews*, 59, 324–334.

Johnstone, R. W. and Paramswaran, A. (2004). *An Introduction to Surface-Micromachining (Information Technology: Transmission, Processing and Storage)*. The Netherlands: Kluwer Academic Publishers, Springer.

Jonsmann, J., Sigmund, O., and Bouwstra, S. (1999). Compliant thermal microactuators. *Sensors and Actuators A-Physical*, 76, 463–469.

Kim, C. and Pisano, A. (1992). Silicon-processed overhanging micro-gripper. *Journal of Microelectromechanical Systems*. Vol. 31–36.

Kim, C. J., Pisano, A. P., Muller, R. S., and Lim, M. G. (1992). Polysilicon microgripper. *Sensors and Actuators A-Physical*, 33, 221–227.

Kim, K., Liu, X. Y., Zhang, Y., and Sun, Y. (2008). Nanonewton force-controlled manipulation of biological cells using a monolithic MEMS microgripper with two-axis force feedback. *Journal of Micromechanics and Microengineering*, 18.

Kim, P. and Lieber, C. M. (1999). Nanotube nanotweezers. *Science,* 286, 2148–2150.

Korzh, V. and Strahle, U. (2002). Marshall Barber and the century of microinjection: From cloning of bacteria to cloning of everything. *Differentiation,* 70, 221–226.

Krijnen, G., Kuijpers, T., Lammerink, T., Wiegerink, R., and Elwenspoek, M. (2003). Comb-drives: Versatile micro-structures for capacitive sensing and electrostatic actuation. In Culshaw, B. and Gobin, P. F. (Eds.) *European Workshop on Smart Structures in Engineering and Technology.* 201–206.

Kuncova, J., Kallio, P., and IEEE (2004). Challenges in capillary pressure microinjection. *Proceedings of the 26th Annual International Conference of the IEEE Engineering in Medicine and Biology Society, Vols 1–7.* 4998–5001

Landau, L. D. and Lifschitz, E.M. (1960). *Mechanics (Their Course of Theoretical Physics),* Oxford, New York: Pergamon Press.

Lee, J. S., Park, D. S. W., Nallani, A. K., Lee, G. S., and Lee, J. B. (2005). Sub-micron metallic electrothermal actuators. *Journal of Micromechanics and Microengineering,* 15, 322–327.

Lee, W. H., Kang, B. H., Oh, Y. S., Stephanou, H., Sanderson, A. C., Skidmore, G., Ellis, M., and IEEE, I. (2003). Micropeg manipulation with a compliant microgripper. *2003 IEEE International Conference on Robotics and Automation, Vols 1–3, Proceedings.*

Legtenberg, R., Groeneveld, A. W., and Elwenspoek, M. (1996). Comb-drive actuators for large displacements. *Journal of Micromechanics and Microengineering,* 6, 320–329.

Li, L. J. and Uttamchandani, D. (2004). Modified asymmetric micro-electrothermal actuator: Analysis and experimentation. *Journal of Micromechanics and Microengineering,* 14, 1734–1741.

Lu, Y. W. and Kim, C. J. (2006). Microhand for biological applications. *Applied Physics Letters,* 89. 164101.

Luo, J. K., Flewitt, A. J., Spearing, S. M., Fleck, N. A., and Milne, W. I. (2005). Comparison of microtweezers based on three lateral thermal actuator configurations. *Journal of Micromechanics and Microengineering,* 15, 1294–1302.

Marie, R., Thaysen, J., Christensen, C. B. V., and Boisen, A. (2005). DNA hybridization detected by cantilever-based sensor with integrated piezoresistive read-out. *Micro Total Analysis Systems 2004,* 2, 485–487.

Molhave, K. and Hansen, O. (2005). Electro-thermally actuated microgrippers with integrated force-feedback. *Journal of Micromechanics and Microengineering,* 15, 1265–1270.

Molhave, K., Hansen, T. M., Madsen, D. N., and Boggild, P. (2004). Towards pick-and-place assembly of nanostructures. *Journal of Nanoscience and Nanotechnology,* 4, 279–282.

Molhave, K., Wich, T., Kortschack, A., and Boggild, P. (2006). Pick-and-place nanomanipulation using microfabricated grippers. *Nanotechnology,* 17, 2434–2441.

Mukundan, V. and Pruitt, B. L. (2009). MEMS electrostatic actuation in conducting biological media. *Journal of Microelectromechanical Systems,* 18, 405–413.

Neagu, C., Jansen, H., Gardeniers, J. G. E., and Elwenspoek, M. (1998). An actuation principle: The electrolysis of water. In Vandenberg, A. and Bergveld, P. (Eds.). *Sensor Technology in the Netherlands: State of the Art.* Dordrecht, NL: Springer. 255–261.

Neild, A., Oberti, S., Beyeler, F., Dual, J., and Nelson, B. J. (2006). A micro-particle positioning technique combining an ultrasonic manipulator and a microgripper. *Journal of Micromechanics and Microengineering,* 16, 1562–1570.

Oh, Y. S., Lee, W. H., Stephanou, H. E., Skidmore, G. D., and Asme (2003). *Design, Optimization, and Experiments of Compliant Microgripper.* ASHE INT. Mechanical Engineering Congress, 2003. Wash, D.C. 345–350.

Ouyang, P. R., Zhang, W. J., Gupta, M. M., and Zhao, W. (2007). Overview of the development of a visual based automated bio-micromanipulation system. *Mechatronics,* 17, 578–588.

Park, D. S. W., Nallani, A. K., Cha, D., Lee, G. S., Kim, M. J., Skidmore, G., Lee, J. B., and Lee, J. S. (2010). A sub-micron metallic electrothermal gripper. *Microsystem Technologies—Micro- and Nanosystems Information Storage and Processing Systems,* 16, 367–373.

Park, J. and Moon, W. (2005). The systematic design and fabrication of a three-chopstick microgripper. *International Journal of Advanced Manufacturing Technology,* 26, 251–261.

Pelesko, J. A. (2002). *Modeling MEMS and NEMS.* Boca Raton, FL: CRC Press.

Reck, K., Richter, J., Hansen, O., Thomsen, E. V., and IEEE (2008). Piezoresistive effect in top-down fabricated silicon nanowires. *MEMS 2008: 21st IEEE International Conference on Micro Electro Mechanical Systems, Technical Digest.* 717–720.

Rubio, W. M., Silva, E. C. N., Bordatchev, E. V., and Zeman, M. J. F. (2009). Topology optimized design, microfabrication and characterization of electro-thermally driven microgripper. *Journal of Intelligent Material Systems and Structures,* 20, 669–681.

Sahu, B., Taylor, C. R., and Leang, K. K. (2010). Emerging challenges of microactuators for nanoscale positioning, assembly, and manipulation. *Journal of Manufacturing Science and Engineering—Transactions of the ASME,* 132. 030917-1–030917-11

Sakaki, K., Dechev, N., Burke, R. D., and Park, E. J. (2009). Development of an autonomous biological cell manipulator with single-cell electroporation and visual servoing capabilities. *IEEE Transactions on Biomedical Engineering,* 56, 2064–2074.

Sardan, O., Andersen, K. N., Macdonald, A. N., Sigmund, O., Boggild, P., Horsewell, A., and ASME (2009). *Focused Ion Beam (Fib) Modification Of Topology Optimized Polysilicon Microgrippers.* DETC 2008: Proc. of ASME Design Engineering Technical conference, New York. Vol. 4. 629–631.

Sardan, O., Eichhorn, V., Petersen, D. H., Fatikow, S., Sigmund, O., and Boggild, P. (2008a). Rapid prototyping of nanotube-based devices using topology-optimized microgrippers. *Nanotechnology,* 19.

Sardan, O., Petersen, D. H., Molhave, K., Sigmund, O., and Boggild, P. (2008b). Topology optimized electrothermal polysilicon microgrippers. *Microelectronic Engineering,* 85, 1096–1099.

Senturia, S. D. (2000). *Microsystem Design.* New York. Springer.

Shimotani, K., Watanabe, H., Manabe, C., Shigematsu, T., and Shimizu, M. (2003). Triple-probe atomic force microscope: Measuring a carbon nanotube/DNA MIS-FET. *Spatially Resolved Characterization of Local Phenomena in Materials and Nanostructures,* 738, 289–292.

Smith, G., Maloney, J., Fan, L., and Devoe, D. L. (2001). Large displacement microactuators in deep reactive ion etched single crystal silicon. In Helvajian, H., Janson, S. W., and Larmer, F. (Eds.). *MEMS Components and Applications for Industry, Automobiles, Aerospace, and Communication.* Proc. of the Society of Photo-optical Instrumentation Engineers (SPIE). San Francisco: Vol. 4559. 138–147.

Stokes, D. (2008). *Principles and Practice of Variable Pressure: Environmental Scanning Electron Microscopy (VP-ESEM).* U.K. John Wiley & Sons.

Sun, Y., Fry, S. N., Potasek, D. P., Bell, D. J., and Nelson, B. J. (2005). Characterizing fruit fly flight behavior using a microforce sensor with a new comb-drive configuration. *Journal of Microelectromechanical Systems,* 14, 4–11.

Tam, E., Lhernould, M. S., Lambert, P., Delchambre, A., and Delplancke-Ogletree, M. P. (2009). Electrostatic forces in micromanipulation: Experimental characterization and simulation including roughness. *Applied Surface Science,* 255, 7898–7904.

Tang, W. C., Nguyen, T. C. H., and Howe, R. T. (1989). Laterally driven polysilicon resonant microstructures. *Sensors and Actuators,* 20, 25–32.

Tanikawa, T. and Arai, T. (1999). Development of a micro-manipulation system having a two-fingered micro-hand. *IEEE Transactions on Robotics and Automation,* 15, 152–162.

Thaysen, J., Boisen, A., Hansen, O., and Bouwstra, S. (2000). Atomic force microscopy probe with piezoresistive read-out and a highly symmetrical Wheatstone bridge arrangement. *Sensors and Actuators A-Physical,* 83, 47–53.

Tombler, T. W., Zhou, C. W., Alexseyev, L., Kong, J., Dai, H. J., Lei, L., Jayanthi, C. S., Tang, M. J., and Wu, S. Y. (2000). Reversible electromechanical characteristics of carbon nanotubes under local-probe manipulation. *Nature,* 405, 769–772.

Umemoto, T., Ayano, K., Suzuki, M., Yasutake, M., Konno, T., and Hashiguchi, G. (2009). Nanotweezers with proximity sensing and gripping force control system. *Japanese Journal of Applied Physics,* 48. 08JB21.

Watanabe, H., Manabe, C., Shigematsu, T., Shimotani, K., and Shimizu, M. (2001). Single molecule DNA device measured with triple-probe atomic force microscope. *Applied Physics Letters,* 79, 2462–2464.

Weck, M. and Peschke, C. (2004). Equipment technology for flexible and automated micro-assembly. *Microsystem Technologies—Micro- and Nanosystems Information Storage and Processing Systems,* 10, 241–246.

Xie, Y., Sun, D., Liu, C., Cheng, S. H., Liu, Y. H., and IEEE (2009). A force control based cell injection approach in a bio-robotics system. *ICRA: 2009 IEEE International Conference on Robotics and Automation, Vols. 1–7.* 2142–2147.

Yan, D., Khajepour, A., and Mansour, R. (2003). Modeling of two-hot-arm horizontal thermal actuator. *Journal of Micromechanics and Microengineering,* 13, 312–322.

Zhang, Y. L., Han, M. L., Shee, C. Y., Ang, W. T., and Chia, T. F. (2007). *Automatic Vision Guided Small Cell Injection: Feature Detection, Positioning, Penetration and Injection.*

Zhou, W. L., Li, W. J. (2004). Micro ICPF actuators for aqueous sensing and manipulation. *Sensors and Actuaturs* A. 114. 406–412.

Zhou, Y. and Nelson, B. J. (1999). Force controlled microgripping. In Nelson, B. J. and Breguet, J. M. (Eds.). *Microrobotics and Microassembly.* Bellingham, SPIE-Int Soc Optical Engineering.

Zyvex. Zyvex Instruments. www.zyvex.com

7 Magnetic Manipulation of Biological Structures

Maria Dimaki and Winnie E. Svendsen
Department of Micro- and Nanotechnology,
Technical University of Denmark, Lyngby

CONTENTS

Magnetic manipulation is a technique that has been used since the middle of the nineteenth century, though in a nonbiological context. At that time, large magnets were used to separate metals from dirt in the mining industry. However, it was not until the invention of what is known as high-gradient magnetic separation (HGMS), usually attributed to J.H.P. Watson (Watson, 1973) that the method has found applications in medicine, chemistry, and biotechnology. The technique allows the trapping of even weakly magnetized particles by producing very strong and very local magnetic field gradients.

Although some biological or chemical species have an inherent magnetic moment, most are essentially nonmagnetic and will therefore not experience any force when placed inside a magnetic field gradient. Therefore, most biological species need to be functionalized with magnetic beads that can then be manipulated by the magnetic field gradients. Today, magnetic beads are commercially available with a variety of coatings, so if one is lucky one only has to choose the correct coated bead and bind it to the biological species in order to be able to manipulate it magnetically.

An example of a test tube magnetic manipulation of beads is illustrated in Figure 7.1. Assume that the solution contains some biological particle (e.g., a protein) and that the magnetic beads are coated with something that binds to this particular protein. After the binding occurs, a magnet can be used to attract the beads and trap them while the solution is changed to another solution, for example, a rinsing buffer or a solution of an antibody that binds to the protein on the bead, and so on.

The same protocol can be used in a microfluidic system, assuming that a structure can be made in the micrometer scale that can produce the high magnetic field gradients required to trap the beads and hold them in place while the fluid is exchanged in the microfluidic channel. This will be discussed later in the chapter. Here, the beads and protein mixture are entered into a microfluidic channel using flow, as illustrated in Figure 7.2. The gray area under the channel is the magnetic microstructure, and the magnetic beads are attached to that once the field is on. Upon binding of the proteins to the beads, the remaining proteins are flushed out of the channel before the field is switched off, and the beads together with the desired proteins are flushed out, too.

The preceding only leaves one question unanswered: Why bother with the effort of an extra step to functionalize the nonmagnetic biological particles with magnetic beads so that we can manipulate them with magnetic fields, instead of just using electrical fields, as described in Chapter 3? It is indeed a relevant question, but there is also a very good answer. In contrast to electrical fields, magnetic fields have minimal effect on the viability of biological matter. Just think about how many people undergo magnetic resonance imaging (MRI) scans every day. This is not the case for electrical fields, which can, for example, cause membrane rupture if they are high enough (Sugar and Neumann, 1984). Moreover, as was described in Chapter 3, the applied voltages inside the liquids create heat that not only can set the liquids in

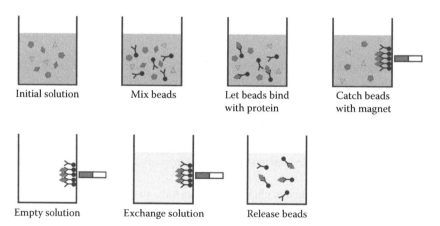

| Initial solution | Mix beads | Let beads bind with protein | Catch beads with magnet |

| Empty solution | Exchange solution | Release beads |

FIGURE 7.1 (**See color insert**) Schematic of the test tube magnetic manipulation of beads.

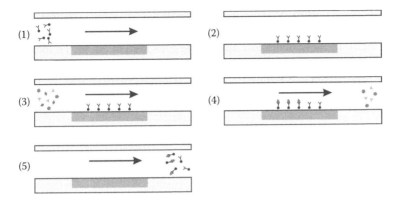

FIGURE 7.2 Schematic of magnetic separation using a microfluidic system. (1) The beads with the bound antibodies are flushed into the channel. (2) The magnetic field is turned on, and the beads are trapped. (3) A collection of different antigens is flushed into the channel. (4) Only the ones fitting to the antibody on the beads bind to it, while the rest are flushed out. (5) The magnetic field is switched off, and the beads together with the bound species are flushed out of the channel.

motion and disrupt the biological particle manipulation but also destroy the biomaterials if the temperature rises too much. Heat generation is not an issue with external magnetic fields, which is certainly an advantage. However, it can be a problem when electromagnets are used on-chip to generate the magnetic field. Finally, while high-conductivity fluids (as most biological buffers are) make electrical field manipulation of particles difficult at best, magnetic manipulation of particles is not affected at all by the chemical properties of the fluid.

In the following sections, we will introduce the magnetic materials, followed by some simple theory on magnetic separation. Finally, some examples of magnetic separation of biological materials from the literature will be given.

7.1 MAGNETIC MATERIALS

In a particle, there are two sources of magnetization: (a) the electrons orbit around the nucleus, as a moving charge creates a magnetic field according to the Biot–Savart law, and (b) the intrinsic magnetic moment or spin of the electrons. Usually, all these magnetic moments in a particle cancel out so that the total magnetic moment is zero. But sometimes, either spontaneously or due to an external magnetic field, these magnetic moments will align and result in a significant magnetic moment of the particle. This ability of the particles is what is used to place them into three distinct categories.

7.1.1 DIAMAGNETIC MATERIALS

Diamagnetic materials are materials that will be magnetized weakly in opposition to an external magnetic field; in other words, they will be repelled by the external field when placed inside it. These materials have a relative magnetic permeability μ_r smaller than 1 and a magnetic susceptibility χ (see Section 7.2), which is negative but small ($\sim 10^{-5}$). In these materials, all the electrons are paired, so the total magnetic moment is zero. This means that they lose their magnetic moment as soon as the external magnetic field is removed. Most materials are diamagnetic, including silver, copper, and gold.

7.1.2 PARAMAGNETIC MATERIALS

These are materials that will be magnetized weakly along the direction of the external magnetic field, so they will be attracted by the external field. They have a relative magnetic permeability that is larger than or equal to 1, and so a positive magnetic susceptibility. In these materials, there are a few unpaired electrons, but the paramagnetic properties will disappear once the external field is removed. Paramagnetic materials are, for example, aluminum, tungsten, lithium, and magnesium.

7.1.3 FERROMAGNETIC MATERIALS

This is perhaps the most interesting category of materials, as they exhibit a strong attraction to external magnetic fields, that is, they have a positive magnetic susceptibility and are able to retain their magnetic properties once the external field is removed. These materials also have unpaired electrons, which have magnetic moments that are aligned with those of their neighbors in order to maintain a lower energy state. Due to energy requirements, the aligned magnetic moments are confined into a domain, which then exhibits a net magnetic moment. A ferromagnetic material, therefore, usually comprises several magnetic domains. When the material is unmagnetized, the magnetic moments of the domains are randomly oriented, and the magnetic moment for the entire material is zero. However, once exposed to an external magnetic field, the domains will align with the field and produce a strong magnetic field. These materials will retain their ferromagnetic properties until they are heated up over a certain temperature called the *Curie temperature*. Ferromagnetic materials include nickel, iron, and cobalt and their alloys.

Another class of ordered magnetic materials with properties similar to ferromagnetic materials is ferrimagnetic materials. The iron oxides as γ–Fe_2O_3 (maghemite) and Fe_3O_4 (magnetite), which are commonly used in magnetic beads, belong to this class.

When dealing with nanoparticles made of ferromagnetic or ferrimagnetic materials, one often meets the term *superparamagnetism*. When a ferromagnet is sufficiently small, it acts similar to a single spin, which can then change moment direction due to thermal agitation. Therefore, if one tries to measure its magnetic moment, it will on average be zero, as for a paramagnet. However, the susceptibility of these materials is much higher than for paramagnets, so they will generate a large magnetization when placed in an external magnetic field. It is due to this phenomenon that there exists a limit for the storage density of computer hard drives.

In the following, ferromagnetic materials will be considered.

7.2 SOME ESSENTIAL DEFINITIONS

Before presenting the equations used for describing how a magnetic particle can be manipulated by an external magnetic field, we give some basic concepts and definitions. When a magnetic material is placed in a magnetic field of strength $\vec{H}$ (in A/m), the individual domain moments align, resulting in the magnetic induction $\vec{B}$ (in tesla), given by

$$\vec{B} = \mu_0(\vec{H} + \vec{M}) \tag{7.1}$$

where μ_0 is the magnetic permeability of free space, and $\vec{M}$ (in A/m) is the magnetization of the particle (i.e., the magnetic moment $\vec{m}$ (in A · m^2) per unit volume). As mentioned earlier, magnetic materials can be categorized in terms of their magnetic susceptibility χ, where

$$\vec{M} = \chi\vec{H} \tag{7.2}$$

Note that Equation 7.2 may give the impression that the relationship between M and H is linear. That is not necessarily the case. The relationship between M and H is plotted in Figure 7.3 for ferromagnetic and superparamagnetic particles (Sung and Rudowicz, 2003; Gijs et al., 2010).

M_{sat} is the saturation value of the magnetization at high magnetic fields, while M_{rem} is the remanent magnetization when the external field is zero. A superparamagnetic particle does not show any significant hysteresis, and the susceptibility in this case is the slope of the linear part (so for low external field strengths) of the curve, as shown in Figure 7.3b.

The susceptibility of the superparamagnetic particle, also known as *effective susceptibility*, is related to the intrinsic susceptibility of the material χ_{mat} as (Gijs et al., 2010)

$$\chi_{eff} = \frac{\chi_{mat}}{1 + N_d\chi_{mat}} \tag{7.3}$$

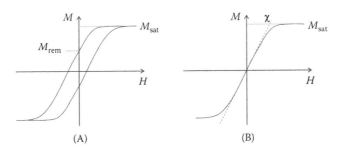

FIGURE 7.3 The relationship between $\vec{M}$ and $\vec{H}$ for (A) a ferromagnetic particle and (B) a superparamagnetic particle.

where N_d is the demagnetization factor, which is equal to 1/3 for spherical particles. Note that the magnetic susceptibility may also depend on the applied magnetic field. It is the effective susceptibility that is featured in Equation 7.2. The demagnetizing effects effectively limit the susceptibility that can be achieved in practice. For a sphere, for example, $\chi_{eff} \approx 3$ when χ_{mat} is very large.

Moreover, by combining Equations 7.1 and 7.2, we find that

$$\vec{B} = \mu_0(1+\chi_{eff})\vec{H} \Rightarrow \begin{cases} \vec{B} = \mu\vec{H} \\ \mu = \mu_0(1+\chi_{eff}) \end{cases} \tag{7.4}$$

where μ is the magnetic permeability of the particle.

7.3 GOVERNING EQUATIONS

In the absence of electric currents or time varying electric fields, the force acting on a magnetic particle of magnetization M in a magnetic field H can be written as (Engel and Friedrichs, 2002)

$$\vec{F} = \mu_0 \oiint d^3r[(\vec{M}\cdot\vec{\nabla})\vec{H}] \tag{7.5}$$

Here, $\vec{H}$ is the magnetic field in the absence of the particle. Equation 7.5 is the general force equation and applies in all cases and for all particle geometries.

7.3.1 NONMAGNETIC MEDIUM

In the case of a nonmagnetic medium, as is the case for all biological buffers, and if the integrant varies slowly across the volume of the particle, Equation 7.5 can be approximated as

$$\vec{F} \approx \mu_0 V_p(\vec{M}\cdot\vec{\nabla})\vec{H} \overset{7.2}{\Rightarrow} F = \mu_0 V_p \chi_{eff}(\vec{H}\cdot\vec{\nabla})\vec{H} \tag{7.6}$$

where V_p is the particle volume, $\vec{M}$ is the magnetization of the particle, $\vec{H}$ is the magnetic field strength, and χ_{eff} is the effective magnetic susceptibility of the particle as given in Equation 7.3. Using a vector identity* and that the field is irrotational ($\vec{\nabla}\times\vec{H} = 0$), this can also be written as

$$\vec{F} = \frac{1}{2}\mu_0 V_p \chi_{eff}\vec{\nabla}\left|\vec{H}\right|^2 \tag{7.7}$$

---

* $\vec{\nabla}(\vec{A}\cdot\vec{B}) = (\vec{A}\cdot\vec{\nabla})\vec{B} + (\vec{B}\cdot\vec{\nabla})\vec{A} + \vec{B}\times(\vec{\nabla}\times\vec{A}) + \vec{A}\times(\vec{\nabla}\times\vec{B})$

This equation can be written in terms of the applied magnetic induction $\vec{B}_{ext}$ by using that $\vec{B}_{ext} = \mu_0 \vec{H}$ (Pankhurst et al., 2003) to obtain

$$\vec{F}_m = \frac{1}{2} V_p \chi_{eff} \frac{1}{\mu_0} \vec{\nabla} |\vec{B}_{ext}|^2 \tag{7.8}$$

Here, $\vec{B}_{ext}$ is the magnetic induction in the absence of the particle. Equation 7.8 is the one commonly encountered in the literature, but in most cases, the symbol $\vec{B}$ is used instead of $\vec{B}_{ext}$. However, this expression can be confusing as $\vec{B}$ is also used to denote the magnetic induction when a particle is magnetized by a field of strength $\vec{H}$. The reader should therefore remember that it is the field (or magnetic induction) *in the absence* of the particle that is featured in the force equation.

7.3.2 NONNEGLIGIBLE MAGNETIC MEDIUM

For a superparamagnetic *spherical* particle of radius R_p in a fluid with nonnegligible magnetic properties and for magnetic fields that vary little over the diameter of the particle, it can be shown (Jones, 1995) that Equation 7.5 results in

$$\vec{F} = 2\pi \mu_m R_p^3 K(\mu_p \cdot \mu_m) \vec{\nabla} |\vec{H}|^2 \tag{7.9}$$

where $K(\mu_p \cdot \mu_m)$ is called the Clausius–Mossotti factor, given in the magnetic case as

$$K(\mu_p \cdot \mu_m) = \frac{\mu_p - \mu_m}{\mu_p + 2\mu_m} \tag{7.10}$$

where μ_p and μ_m are the intrinsic magnetic permeabilities of the particle and the medium, respectively. Equation 7.9 can be rewritten in terms of the susceptibility as

$$\vec{F} = 2\pi \mu_0 (1 + \chi_m) R_p^3 \frac{\chi_p - \chi_m}{3 + \chi_p + 2\chi_m} \vec{\nabla} |\vec{H}|^2 \tag{7.11}$$

Note that as the definition of the Clausius–Mossotti factor includes the demagnetization effects, the susceptibilities in Equation 7.11 are not the effective susceptibilities but the intrinsic material susceptibilities. It can be shown that, at the limit of a very large χ_p, Equation 7.11 results in Equation 7.7 for $\chi_{eff} = 3$.

Note that in special cases, where χ_m is quite close to zero and χ_p is on the order of 1, Equation 7.11 can be written as

$$\vec{F} \approx \frac{1}{2} \mu_0 V_p \nabla \chi \vec{\nabla} |\vec{H}|^2 \tag{7.12}$$

with $\nabla \chi = \chi_p - \chi_m$.

This equation, or alternatively expressed in terms of the applied magnetic induction, is often encountered in the literature as the magnetic force on a particle, but it is important to remember that it is an approximation.

For most practical purposes, the magnetization of the fluid is negligible, and Equation 7.7 can be used. The effective particle susceptibility is usually measured or given by the manufacturer.

7.3.3 VISCOUS AND GRAVITATIONAL FORCES

Given that the particle is in a moving fluid, there are two more forces to consider: the viscous drag and the gravitational force. The viscous drag is given by

$$\overrightarrow{F_d} = 6\pi\eta R_p(\vec{u} - \vec{v}) \tag{7.13}$$

where η is the viscosity of the fluid, $\vec{u}$ is the fluid velocity, and $\vec{v}$ is the particle velocity. The gravitational force is given by

$$\overrightarrow{F_g} = V_p(\rho_b - \rho_f)\vec{g} \tag{7.14}$$

where ρ_b and ρ_f are the particle and fluid densities, respectively, and $\vec{g}$ is the gravitational acceleration.

7.3.4 PARTICLE MOTION

Using Newton's second law, the particle's velocity can be calculated:

$$\overrightarrow{F_{tot}} = \overrightarrow{F_m} + \overrightarrow{F_g} + \overrightarrow{F_d} = m\frac{d\vec{v}}{dt} \tag{7.15}$$

This equation is the same one used in Chapter 3 for the calculation of the particle velocity in the case of a dielectrophoretic force acting on the particle. Following the same arguments as in Chapter 3, the particle velocity can be written as

$$\vec{v} = \frac{2R_p^2}{9\eta}\left(\frac{\chi_{\text{eff}}}{2\mu_0}\vec{\nabla}|B_{ext}|^2 + (\rho_b - \rho_f)\vec{g}\right) + \vec{u} \tag{7.16}$$

Although the magnetic beads are small, up to a few micrometers in diameter, their density is large and the gravitational forces on them are not insignificant. However, when the gravitational forces are much smaller than the magnetic forces, Equation 7.16 becomes

$$\vec{v} = \frac{\chi_{\text{eff}}R_p^2}{9\mu_0\eta}\vec{\nabla}|B_{ext}|^2 + \vec{u} \tag{7.17}$$

From Equation 7.17, the so-called magnetophoretic mobility can be defined as

$$\xi = \frac{\chi_{\text{eff}} R_p^2}{9\eta} \tag{7.18}$$

This parameter is a particle property and is a measure of the ability of the particle to be manipulated by a magnetic field in a certain fluid.

Note that the magnetophoretic mobility is sometimes defined as

$$\xi = \frac{2\nabla\chi R_p^2}{9\eta} = \frac{V\nabla\chi}{6\pi\eta R_p} \tag{7.19}$$

depending on whether or not the vector identity has been used to simplify Equation 7.6 combined with the approximation of Equation 7.12.

7.4 INTRODUCTION TO MAGNETIC BIOSEPARATION

In this section we will mainly discuss manipulation using magnetic beads, which clearly dominates the scene, but we will also briefly mention the use of the magnetic properties of the biomaterial on its own.

Magnetic bioseparation has a long trace back in history; an early review of the application of magnetic separation in biotechnology is given in the paper from Christine Setchell already in 1985 (Setchell, 1985). Several reviews have later emerged on the topic of magnetic beads or cell handling in microsystems in some fashion (e.g., Pankhurst et al., 2003; Pamme et al., 2006; Gijs, 2004); for a recent and detailed overview of magnetic particles used in microfluidics applications on biological analysis, the review by Gijs et al. (2010) is worth reading. In this section, the relevant topics from these reviews will be summarized together with the development of the magnetic systems over the years; here, the papers of Smistrup proved useful (Smistrup, 2007; Bu et al., 2008; Smistrup et al., 2008; Bu et al., 2007; Smistrup et al., 2007; Smistrup et al., 2006a; Smistrup et al., 2006b; Smistrup et al., 2005b; Smistrup et al., 2005c; Smistrup et al., 2005a).

Using magnetic manipulation, microsystems gained momentum with the development of the high-gradient magnetic separation (HGMS) systems in the early 1970s, which enabled the relatively simple generation of strong magnetic gradients (Oberteuffer, 1973; Watson, 1973; Zborowski et al., 2003; Safarik and Safarikova, 1999; 2004).

One of the major advantages of using magnetic separation techniques in contrast to, for example, electric techniques is that magnetic interactions are generally not affected by surface charges, pH, ionic concentrations, or temperatures, which are often very important parameters to keep track of in biological systems.

The simplest way to manipulate magnetic particles, a method used in the early experimental days, is to hold a magnet close to the substrate, as illustrated in Figure 7.4, where magnetic particles (DYNA beads) were attached to peptide nanostructures and then manipulated using an external permanent magnet. This somewhat crude

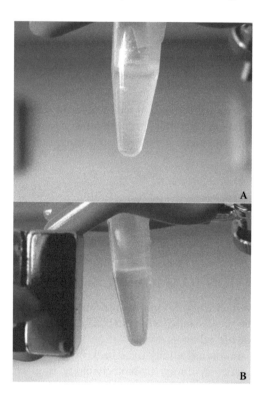

FIGURE 7.4 (A) Peptide nanostructures functionalized with magnetic beads. (B) When an external magnetic field is applied, all the peptide structures move toward the magnet and separate from the rest of the solution.

method was also applied to microsystems at first; magnetic beads would enter the fluidic system, and a larger permanent magnet was then used to trap the magnetic particles attached to the biological species of interest. While this is sufficient for some applications, we will discuss more sophisticated methods to generate the magnetic field and integrate this with the microfluidic systems. Special emphasis will be given to integrated magnetic manipulation systems combined with microfluidic devices. Combined, these two techniques give powerful manipulation opportunities for biological samples. Microfluidic theory is addressed separately in Chapter 8.

7.5 MAGNETIC MANIPULATION SYSTEMS

In general, when talking about magnetic manipulation in microsystems, the magnetic systems can be divided into "active" and "passive." Active systems refer to systems *with integrated electromagnetic* structures. Passive systems are here categorized as systems with *permanent magnets* or with *soft magnetic structures* magnetized by an external magnetic field integrated in the microdevice.

Both scenarios will be described shortly below with an emphasis on how to make the magnetic fields or systems on microscale. As discussed in Section 7.4, the simple method of using permanent magnets at the wall of the area of interest is not optimal. It is preferable to increase the manipulator efficiency, for example, by producing regions of high and low magnetic field gradient, where the magnetic particle can be manipulated in the high field gradients.

In this respect, it would be appropriate here to introduce the term *magnetic cell sorting* (MACS). The term was introduced by Milternyi Biotec in 1990 and is used for cell sorting in a manner similar to fluorescent-assisted cell sorting (FACS). The MACS system can be implemented in a number of different ways, but mainly passive magnetization combined with HGMS is used. The methods will be discussed in Section 7.5.2.

7.5.1 ACTIVE SYSTEMS

Active systems are microsystems with integrated electromagnetic structures for generating the magnetic field—possibly combined with an external field. The field can be generated by current lines or microfabricated electromagnets. It is important to mention that the major difference between using a permanent magnet and an electromagnet is the much lower fields generated in the electromagnetic systems. A permanent magnet easily generates a magnetic induction of 0.5–1 T, whereas an electromagnet magnetic induction is in the order of millitesla.

An early example of such a system was presented by Chong H. Ahn and Mark G. Allen in 1994. In this paper, they used copper coils to supply the magnetic field along the microfluidic channel. A field of 0.03 T could be achieved in the microfluidic channel. They suggested that the device could be used as, for example, a magnetic microsensor, as microactuator or micromagnetic power device. This group continued to work on new designs to optimize the magnetic fields in the devices; they also published one of the first total analysis systems using magnetic manipulation, including an electrochemical sensor with electric readout (Ahn et al., 1996; Choi et al., 2000; Choi et al., 2001b; Choi et al., 2001c; Choi et al., 2002). Later, they took the magnet off-chip to reduce fabrication cost and to achieve larger magnetic fields (Do et al., 2004). The Ahn group took several other routes to optimize magnetic fields for use in microfluidics (Liakopoulos et al., 1997; Choi et al., 2001a; Rong et al., 2003). Several other designs of microfabricated electromagnets were developed in the coming decade. Ramadan designed a system, where the aim was to get the magnets as close to the channel as possible. His designs were small compared to the spiral designs of Ahns group. He used material both with and without magnetic cores. With these systems, they showed that they were able to both capture and control the movement of magnetic beads (Ramadan et al., 2004; Ramadan et al., 2006a; Ramadan et al., 2006c; Ramadan et al., 2006b). Early work on manipulation of cells using magnetic actuation was demonstrated by Westervelt's group. They made an array of magnetic bead traps composed of planar on-chip microelectromagnets, which were individually addressable, enabling transport of the beads inside a microfluidic system (Lee et al., 2001; Lee et al., 2004a; Lee et al., 2004b; Lee et al., 2005).

Roel Wirix-Speetjens has also investigated capture and transport of magnetic beads using tapered conductors inside microfluidic channels. They were able to transport beads with speeds up to 60 μm/s (Wirix-Speetjens and de Boeck, 2004; Wirix-Speetjens et al., 2005; Wirix-Speetjens et al., 2006).

The foregoing active systems above all include some form of silicon microfabrication, and, in general, they are expensive to fabricate. Adam C. Siegel of the Whitesides group recently demonstrated that on-chip electromagnets and microfluidic channels could be fabricated in PDMS using solder as the current line in a two-step process (Siegel et al., 2006). By applying a current of 1 A through the current line, they were able to move 90% of the magnetic beads to one side of the 100 μm wide microfluidic channel within 1 s.

Active magnetic systems have the major advantage that the coils and current lines can be adjusted and changed in a device, leaving some flexibility. Also, the fact that they can be individually addressed leads to several specific applications for manipulation. The drawbacks are the low magnetic field achievable in the microfluidic channel (~30 mT), they are difficult and expensive to fabricate, and also they may generate heat very close to the microfluidic channel, which may kill biological entities in the channel.

7.5.2 Passive Systems

In this section, passive magnetic systems are reviewed. Such systems are characterized by permanent magnetic structures. The most interesting topic here is to integrate a magnetic structure on the chip. The most common method is to integrate a soft magnetic material in the microdevice, which then is magnetized by large external magnetic fields.

While passive systems are less common in literature compared to the active systems described earlier, they have the potential of being much more effective than their active counterparts mainly due to the higher fields that can be generated, but also because—and maybe more importantly for biomanipulation—no extra heat is generated on the chip. Also, if the external field is large enough to magnetically saturate the magnetic beads, then the job of maximizing the force is equal to maximizing the magnetic field gradient. Soft magnetic materials generally have saturation fields on the order of 1 T, and if the geometry can be designed such that the characteristic length scale for the magnetic field is on the order of a 100 μm or less, then magnetic field gradients in the range of 10^4 T/m are well within reach.

As mentioned earlier, the principle of magnetic cell sorting (MACS) was introduced in 1990 by Milternyi Biotec. In MACS, cells are bound to superparamagnetic beads and then retained in a high-gradient magnetic field generated by a magnetic column in the field of external magnets. The sample is flushed through the magnetic column; labeled cells are trapped in the column, and unlabeled cells will pass through the column without being affected by the magnetic field. Later, the labeled cells can be collected by turning off the magnetic field and be further analyzed.

The use of a quadruple magnetic configuration along a fluid channel was proposed in 1998. This geometry enabled manipulation by adjusting the magnetic field gradients (Chalmers et al., 1998; Hatch and Stelter, 2001; Williams et al., 1999). Using a permanent dipole configuration was also investigated and with some success. Devices have been made for separation using magnetic beads, and even cell tracking velocities were measured (Nakamura et al., 2001b; Zhang et al., 2005; Moore et al., 2000; Blankenstein and Larsen, 1998; Pamme and Manz, 2004; Nandy et al., 2008).

An alternative passive system is to integrate micropatterned soft material into the field generated by a larger magnet, much similar to the MACS principle but in the microscale. Whitesides published in 2002 (Deng et al., 2002) a microfluidic system containing 7 μm high cylindrical nickel posts with a diameter of 15 μm. The posts were then magnetized with a neodymium iron boron (NdFeB) permanent magnet and used as a sorting system. The system is reminiscent of the HGMS system devised by Watson in 1973 (Watson, 1973). For biological manipulation, the drawback of this design is that the sample is in touch with the nickel.

Several other methods of incorporating magnetizable material in microsystems were investigated in this period. In 2005, Ichikawa and co-workers demonstrated a system with patterned sputtered permalloy film beneath a microfluidic channel (Ichikawa et al., 2005). The film had a thickness of only 500 nm, which meant that forces from the film were short ranged, but retention of a plug of beads was successfully demonstrated. Examples of other systems based on paramagnetic materials have been proposed by Mirowski et al. (2004), Mirowski et al. (2005), Moreland et al. (2005), Inglis et al. (2004, 2006), Nedelcu and Watson (2002), Hoffmann and Franzreb (2004), Todd et al. (2001), and Haik et al. (1999).

In 2004 and 2006, Han and Frazier demonstrated a system for continuous separation of red and white blood cells based on their intrinsic magnetic properties (red blood cells are paramagnetic and white blood cells are diamagnetic) (Han and Frazier, 2004; Han and Frazier, 2006). The system consists of a nickel wire placed in the middle of a microfluidic channel, and this wire either repels or attracts paramagnetic cells depending on the direction of the applied magnetic field.

Inglis et al. (2004) presented a system with nickel stripes at the bottom of a microfluidic channel. The nickel stripes were placed at an angle to the fluid flow, and magnetic particles would tend to follow the direction of the magnetic stripes, and would thus be deflected and separated from the rest of the sample. Inglis and co-workers did a review on microfluidic systems for magnetic cell separation in 2006 (Inglis et al., 2006).

A second class of microfluidic passive magnetic separators exists. These all use external permanent magnets, and depend on the fact that in microfluidic systems, fluid flow is laminar, and thus coflowing fluids do not readily mix. The principle is that of magnetic field flow fractionation, which was described in 1999 by Zborowski et al. (1999). Magnetic field flow fractionation (MFFF) is an interesting method in which the combination of magnetic fields and microfluidic are both used actively in the sorting process. In this method, a magnetic force acts on the magnetic beads together with hydrodynamic forces, enabling sorting of different magnetic bead in the device (Giddings et al., 1976).

Gert Blankenstein and Ulrik Darling Larsen demonstrated this principle in a microfluidic system in 1998 (Blankenstein and Larsen, 1998). In a channel with two inlets, sample was infused in one inlet, and buffer was infused in the other. The two streams do not mix due to the laminar flow, but by application of a magnetic field, magnetic particles were pulled from the sample stream into the buffer stream. The two fluids were split into two outlets downstream.

Steen Ostergaard of the same group demonstrated a slightly more advanced system based on the same approach in 1999 (Ostergaard et al., 1999). Kim and Park applied the same principle for a multiplexed immunoassay in 2005 (Kim and Park, 2005).

Pamme and co-workers published several experimental studies of a system similar to that described by Blankenstein and Larsen (1998). However, their system had the advantage of multiple outlets. The systems in the preceding text usually only have two outlets, which means that particles are either measured as magnetic or nonmagnetic. With multiple outlets, Pamme and co-workers were able to separate several kinds of particles based on the induced magnetic moment of the particles, that is, particles with higher susceptibility are separated from particles with lower susceptibility, and smaller particles can be separated from larger particles (Pamme and Manz, 2004; Pamme et al., 2006; Pamme and Wilhelm, 2006).

The group of Nan Xia presented a system in 2006 for continuous separation of magnetic beads from nonmagnetic beads and *E. coli* bacteria bound to magnetic beads from a sample using a magnetic structure (Xia et al., 2006).

Torsten Lund-Olesen demonstrated how the efficiency of passive magnetic separators could be enhanced by the addition of herringbone mixer structures in the bottom of the microfluidic channel (Lund-Olesen et al., 2006; Lund-Olesen et al., 2007).

7.5.3 HYBRID SYSTEMS

Hybrid systems are here described as having on-chip current lines or coils together with a homogenous external magnetic field. The on-chip current lines generate the magnetic field gradients, and the external applied field magnetizes the magnetic beads. By combining active and passive systems in this fashion, you obtain the addressability and flexibility of the active system and the strong magnetic fields of the passive system. Furthermore, an added feature of switching from attractive to repulsive force is possible; due to the fact that when the current is turned on, the field from the conductor will couple constructively with the external field in some positions and destructively in other positions. When the current in the conductor is reversed, these positions are switched and the opposite effect is achieved. Neither passive nor active systems can do that.

Tao Deng of the Whitesides group published a system in 2001 with two serpentine conductors next to each other. By switching the current in the two conductors, they could make magnetic beads move along the conductors (Deng et al., 2001). Rida of the Gijs group presented a system capable of doing the same thing, but where the Deng system would move beads at speeds ≈100 μm/s, the Rida system was capable of a speed of ≈400 μm/s. The Rida system uses overlapping coils on a printed circuit board to supply the moving force (Rida et al., 2003). Marc Tondra, Nikola Pekas, and co-workers have presented systems where current lines are buried beneath the bottom of a microfluidic channel. The magnetic field from the buried current lines coupled with an external field and would divert magnetic beads in the flow to either of two channel outlets (Tondra et al., 2001; Pekas et al., 2005).

7.6 MAGNETIC BEAD MANIPULATION WITH BIOLOGICAL FUNCTIONALITIES

Labeled detection, for example, monitoring a biological reaction by labeling one or several components in the reaction, is by now standard techniques. Labels used span from fluorescent molecules, quantum dots, nanoparticles, radio isotope, enzymes, and charged molecules to magnetic beads. Magnetic beads, which is the topic of discussion here, have many advantages compared to the other labels mentioned. The magnetic properties are stable over time; the magnetism is not altered by temperature, pressure, or chemical reagents. Furthermore, they are generally no magnetic noise or interference; neither are the fields screened in aqueous solutions or reagents.

Magnetic bead labeling as a manipulation technique in microsystems has been reviewed by several authors (Gijs et al., 2010; Megens and Prins, 2005; Graham et al., 2004; Tamanaha et al., 2008). Magnetic bead manipulating can, of course, also be as simple as the addition or removal of an external magnet to a test tube as illustrated in Figures 7.1 and 7.4; however, in this section, we will concentrate on magnetic bead separation in microfluidic devices, where the microfluidics and the magnetic properties combined give unique manipulation possibilities.

Looking at the theory developed in Section 7.3, you can utilize the spatial and temporal combination of the forces and thus design procedures for manipulation protocols. In Figure 7.2, the principle behind magnetic bead manipulation in a microfluidic system is illustrated very nicely. Here, a schematic of an immunoassay is illustrated. The magnetic beads containing the antibody are flowing in a microfluidic device. When encountering the magnetic field, the labeled beads are retained from a flow, and the beads are fixed at the bottom of the channel. The next step is to send the solution, for example, blood to be analyzed into the microchannel. If antigens exist in the solution, they will be trapped by the antibody on the beads. The channel can then be rinsed and the immunoassay released for further analysis. The same scenario could be utilized as a purification process, where the impurities to extract are labeled with magnetic beads. These types of experiments are well described in the recent review by Gijs (Gijs et al., 2010), and some more examples will be given later in this chapter.

To conduct this kind of experiment on cells, the magnetophoretic mobility of a magnetically tagged cell is important to calculate for precise control. This is the parameter that makes it possible to separate magnetically labeled cells from unlabeled cells. The magnetophoretic mobility for a magnetic bead that we will use in this analysis is defined in Equation 7.19. The situation for an immunomagnetically labeled cell as described in the preceding example is far more complex.

The magnetic force will depend on the number of magnetic particles attached on the cell membrane, and the viscous drag force in the liquid of viscosity η should account for the whole cell–bead complex.

Therefore, in the case of a two-step binding protocol as illustrated in Figure 7.2 and described in Gijs et al. 2010, Section 5.2 (Gijs et al., 2010), it is necessary to rewrite Equation 7.19 for a single magnetic bead as described in Section 7.3.4. Doing this, it turns out (Gijs et al, 2010) that five parameters will influence the magnetophoretic mobility of an immunomagnetically labeled cell. These are repeated here for ease of reading and an example of how they are used can be found in Gijs et al, 2010.

The five parameters are: the antibody binding capacity (ABC) of a cell population, the secondary antibody binding amplification (ψ), the number of magnetic beads bound to one antibody (n), the particle magnetic field interaction parameter as in the original equation ($V\Delta\chi$), and the cell diameter (D_c). Putting this together we can rewrite Equation 7.19 as follows:

$$\xi_{cell} = \frac{ABC\psi n V\nabla\chi}{3\pi D_c \eta} \tag{7.20}$$

Several studies has been undertaken to understand the influence of these parameters, and more information can be found in the review by Gijs et al. (Gijs et al., 2010) and references therein.

7.7 MAGNETIC CELL MANIPULATION

Direct magnetic cell manipulation is not as common as using magnetic beads for manipulation; however, some biomolecules do exhibit intrinsic magnetic fields and can thus be separated or manipulated directly. The most well-known example of cells exhibiting intrinsic magnetic fields are deoxygenated red blood cells (RBC). Hemoglobin is the iron-containing protein in RBC; oxygenated hemoglobin is dia-magnetic due to the presence of paired electrons on the ion atoms and has therefore very low magnetic susceptibility (Table 7.1). With fields in the tesla range, this dif-ference in magnetic susceptibility between water and the RBC is enough to achieve separation. Other biological compounds such as the magnetotactic bacteria form a type of naturally magnetic cells (Gijs et al., 2010). Although possible, it is often not convenient to use the cells' intrinsic magnetic properties for manipulation, mainly due to the high magnetic fields required to do so.

TABLE 7.1
Difference between the Susceptibility of Oxygenated and Deoxygenated Hemoglobin and that of Water

	Magnetic Susceptibility
χ_{water}	-9.0×10^{-6}
$\Delta\chi_{oxy}$	
($\chi_{RBC,oxy}-\chi_{water}$)	-0.19×10^{-6}
$\Delta\chi_{deoxy}$	
($\chi_{RBC,deoxy}-\chi_{water}$)	3.3×10^{-6}

Note: The values for the magnetic susceptibility is taken from Spees (Spees, W. M., Yablonskiy, D. A., Oswood, M. C., and Ackerman, J. J. H. *Magnetic Resonance in Medicine*, 45, 533–542, 2001) and Zborowski (Zborowski, M., Ostera, G. R., Moore, L. R., Milliron, S., Chalmers, J. J., and Schechter, A. N. *Biophysical Journal*, 84, 2638–2645, 2003).

7.8   MAGNETIC CELL SEPARATION

Cell separation has been a major driving force in the evolution of manipulation techniques. It is the dream of any doctor to pinpoint unhealthy cells fast, accurately, and specifically. A lot of different microfluidic techniques promise to fulfill this dream. Several techniques have been tried, for example, optical, mechanical, gravitational, chemical, fluidic, and magnetic methods, to achieve this goal all with some successes, but no universal method is found. In this section, we will concentrate on magnetic separation techniques for cell separations. The most widely used method to separate and sort cells today is to use the fluorescence-activated cells sorting (FACS), first described by Bonner et al. in 1972 (Bonner et al., 1972). The magnetic counterpart, magnetic cell sorting (MACS) is an alternative, since it has some advantages. In MACS, a permanent magnet generates a strong magnetic gradient in a magnetic column, and then, cells labeled by superparamagnetic nanoparticles are passed through this column and separated from unlabeled cells. The MACS method of separation is less expensive than FACS, but MACS cannot yield as much information as in FACS and is therefore often used as a preparation before further analysis, for example, using FACS. As discussed in Section 7.5.1, generating high magnetic field gradients in microdevices is possible. This led to the idea of combining MACS and microfluidics, which could have several advantages so that sorting and analysis could be carried out in one step, for example, point of care analysis (Gijs et al., 2010).

In this section, a summary of using magnetic manipulation on cells is given. The examples are taken from the very new review by Gijs et al. in 2010 (Gijs et al., 2010).

Separation of red blood cells using magnetic properties was first demonstrated in a *Nature* paper by Melville et al. (1975). Here, a column packed with magnetic wires was placed in a magnetic induction of 1.75 T, yielding a gradient close to the wire of approximately $8 \cdot 10^3$ T/m. Whole blood mixture was introduced into the column, the field was turned off, and the RBC were eluded from the column. Other devices have since been developed using this principle (Graham, 1981; Melville et al., 1982; Takayasu et al., 2000; Paul et al., 1981a; Paul et al., 1981b; Fuh et al., 2004). Combining this sorting method with microfluidics provides multiple functionalities in one system, for example, by the separation of oxygenated and deoxygenated RBCs (Zborowski et al., 2003), or continuous separation of RBCs from whole or diluted blood (Han and Frazier, 2005; Jung and Han, 2008; Qu et al., 2008). By constructing specific magnetic gradients in a microdevice, it was shown that it was possible to obtain the separation of WBC from RBC; 93% WBC and 97% RBC were separated from whole blood, continuously at a rate of 5 µL/h (Han and Frazier, 2006). It was later demonstrated that nucleated RBCs could be separated from maternal blood samples enabling monitoring and diagnostic of maternal, fetal, and neonatal diseases (Huang et al., 2008).

Separation of white blood cells presents some challenges due to the low concentration in whole blood (~5 – 10000/µL). Since analysis of WBCs can indicate several serious diseases—for example, the ratio of CD4/CD8 cells, which are subtypes of WBC, is a measure of how the human immunodeficiency virus (HIV) progresses—it is of great interest to construct very sensitive and precise separation methods for WBCs. In 2004, Inglis et al. (2004) demonstrated an efficient extraction of human

WBC from whole blood using continuous flow in a microfluidic device. Embedded in the device were tiny magnetic strips that were magnetized by an external applied filed of only 0.8 T. The cells were magnetically labeled using anti CD45 magnetic beads; these were attracted to the stripes and followed the direction of these, whereas the unlabeled followed the fluid flow direction.

Similar systems were later developed for different types of WBCs; several examples of WBC manipulation are listed in Gijs et al. (2010). Issues often encountered when combining microdevices with magnetic manipulation of biomaterial are adhesion of the biomaterial to the microchannel walls. A droplet-based magnetic separator has been developed to avoid this and shows efficiency of cell separation comparable to that of current-state commercial systems (Kim et al., 2007).

Separation of stem cells is a challenging task; isolation of very rare hematopoietic progenitor cells such as stem cells from the human umbilical cord is very difficult, and more and more needed. Magnetic cell separation techniques such as MACS column have been utilized with good results (Powles et al., 2000; Weissman, 2000).

Separation of cancer cells from blood samples is a very fast growing field. For example, the presence of circulating tumor cells (CTCS) in the blood stream is an indicator of metastatic disease in cancer patients. The concentration of these cells is a vital information in terms of prognosis and treatment of several cancer types, for example, in breast cancer (Cristofanilli et al., 2004). A count of one CTC cell per milliliter or less in whole blood is an indicator of lower risk of relapse following chemotherapy. Several batch-type commercial systems have been used for separation of these markers both with fluorescent (FACS) and magnetic labeling using nanoparticles (MACS). However, some issues arise similar to trapping of cells in the magnetic column (Comella et al., 2001). Continuous flow magnetic separation systems show promising results in the isolation of rare cancer cells from blood samples, for example, human breast cancer cells MCF-7 were labeled and detected in Chalmers et al. 1999a, 1999b and, HCC1954 and MCF-7 have been used as examples in various magnetic separation setups (e.g., quadruple magnetic flow sorter) (Nakamura et al., 2001a; Yang et al., 2009).

Immunomagnetic labeling has also been tried for human cervical cancer cells (HeLa) (Pamme and Wilhelm, 2006). The nickel micropillar system has also been tested enabling capture of A549, a human lung carcinoma cells using magnetic beads as labels (Liu et al., 2007).

Separation of bacteria cells using magnetism can be useful in, for example, rapid detection of microbial contamination in different areas of interest (e.g., industrial water lines, food sector, clinical microbiology). Systems developed using paramagnetic particles incubated with the bacteria, where the bacteria engulf the nanoparticles and can hence be separated in a magnetic separation system, are described in (Zborowski et al., 1993a; Zborowski et al., 1993b).

REFERENCES

Miltenyi Biotec. www.miltenyibiotec.com

Ahn, C. H. and Allen, M. G. (1994). Fully integrated micromachined magnetic particle manipulator and separator. *Proceedings of the IEEE Micro Electro Mechanical Systems*, 91–96.

Ahn, C. H., Allen, M. G., Trimmer, W., Jun, Y. N., and Erramilli, S. (1996). A fully integrated micromachined magnetic particle separator. *Journal of Microelectromechanical Systems,* 5, 151–158.

Blankenstein, G. and Larsen, U. D. (1998). Modular concept of a laboratory on a chip for chemical and biochemical analysis. *Biosensors and Bioelectronics,* 13, 427–438.

Bonner, W. A., Sweet, R. G., Hulett, H. R., and Herzenbe.La (1972). Fluorescence activated cell sorting. *Review of Scientific Instruments,* 43, 404–&.

Bu, M., Christensen, T. B., Smistrup, K., Wolff, A., and Hansen, M. F. (2007). A High-throughput SU-8 microfluidic magnetic bead separator. *Transducers '07 and Eurosensors Xxi, Digest of Technical Papers, Vols. 1 and 2.*

Bu, M. Q., Christensen, T. B., Smistrup, K., Wolff, A., and Hansen, M. F. (2008). Characterization of a microfluidic magnetic bead separator for high-throughput applications. *Sensors and Actuators A-Physical,* 145, 430–436.

Chalmers, J. J., Haam, S., Zhao, Y., McCloskey, K., Moore, L., Zborowski, M., and Williams, P. S. (1999a). Quantification of cellular properties from external fields and resulting induced velocity: Cellular hydrodynamic diameter. *Biotechnology and Bioengineering,* 64, 509–518.

Chalmers, J. J., Haam, S., Zhao, Y., McCloskey, K., Moore, L., Zborowski, M., and Williams, P. S. (1999b). Quantification of cellular properties from external fields and resulting induced velocity: Magnetic susceptibility. *Biotechnology and Bioengineering,* 64, 519–526.

Chalmers, J. J., Zborowski, M., Sun, L. P., and Moore, L. (1998). Flow through, immunomagnetic cell separation. *Biotechnology Progress,* 14, 141–148.

Choi, J. W., Ahn, C. H., Bhansali, S., and Henderson, H. T. (2000). A new magnetic bead-based, filterless bio-separator with planar electromagnet surfaces for integrated bio-detection systems. *Sensors and Actuators B-Chemical,* 68, 34–39.

Choi, J. W., Liakopoulos, T. M., and Ahn, C. H. (2001a). An on-chip magnetic bead separator using spiral electromagnets with semi-encapsulated permalloy. *Biosensors and Bioelectronics,* 16, 409–416.

Choi, J. W., Oh, K. W., Han, A., Okulan, N., Ajith Wijayawardhana, C., Lannes, C., Bhansali, S., Schlueter, K. T., Heineman, W. R., Halsall, H. B., Nevin, J. H., Helmicki, A. J., Thurman Henderson, H., and Ahn, C. H. (2001b). Development and characterization of microfluidic devices and systems for magnetic bead-based biochemical detection. *Biomedical Microdevices,* 3, 191–200.

Choi, J. W., Oh, K. W., Thomas, J. H., Heineman, W. R., Halsall, H. B., Nevin, J. H., Helmicki, A. J., Henderson, H. T., and Ahn, C. H. (2001c). An integrated microfluidic biochemical detection system with magnetic bead-based sampling and analysis capabilities. *14th IEEE International Conference on Micro Electro Mechanical Systems, 2001. Interlaken Technical Digest,* 447–450.

Choi, J. W., Oh, K. W., Thomas, J. H., Heineman, W. R., Halsall, H. B., Nevin, J. H., Helmicki, A. J., Henderson, H. T., and Ahn, C. H. (2002). An integrated microfluidic biochemical detection system for protein analysis with magnetic bead-based sampling capabilities. *Lab on a Chip,* 2, 27–30.

Comella, K., Nakamura, M., Melnik, K., Chosy, J., Zborowski, M., Cooper, M. A., Fehniger, T. A., Caligiuri, M. A., and Chalmers, J. J. (2001). Effects of antibody concentration on the separation of human natural killer cells in a commercial immunomagnetic separation system. *Cytometry,* 45, 285–293.

Cristofanilli, M., Budd, G. T., Ellis, M. J., Stopeck, A., Matera, J., Miller, M. C., Reuben, J. M., Doyle, G. V., Allard, W. J., Terstappen, L., and Hayes, D. F. (2004). Circulating tumor cells, disease progression, and survival in metastatic breast cancer. *New England Journal of Medicine,* 351, 781–791.

Deng, T., Prentiss, M., and Whitesides, G. M. (2002). Fabrication of magnetic microfiltration systems using soft lithography. *Applied Physics Letters,* 80, 461–463.

Deng, T., Whitesides, G. M., Radhakrishnan, M., Zabow, G., and Prentiss, M. (2001). Manipulation of magnetic microbeads in suspension using micromagnetic systems fabricated with soft lithography. *Applied Physics Letters*, 78, 1775–1777.

Do, J., Choi, J. W., and Ahn, C. H. (2004). Low-cost magnetic interdigitated array on a plastic wafer. *IEEE Transactions on Magnetics*, 40, 3009–3011.

Engel, A. and Friedrichs, R. (2002). On the electromagnetic force on a polarizable body. *American Journal of Physics*, 70, 428–432.

Fuh, C. B., Su, Y. S., and Tsai, H. Y. (2004). Determination of magnetic susceptibility of various ion-labeled red blood cells by means of analytical magnetapheresis. *Journal of Chromatography A*, 1027, 289–296.

Giddings, J. C., Yang, F. J. F., and Myers, M. N. (1976). Flow field-flow fractionation—versatile new separation method. *Science*, 193, 1244–1245.

Gijs, M. A. M. (2004). Magnetic bead handling on-chip: New opportunities for analytical applications. *Microfluidics and Nanofluidics*, 1, 22–40.

Gijs, M. A. M., Lacharme, F., and Lehmann, U. (2010). Microfluidic applications of magnetic particles for biological analysis and catalysis. *Chemical Reviews*, 110, 1518–1563.

Graham, D. L., Ferreira, H. A., and Freitas, P. P. (2004). Magnetoresistive-based biosensors and biochips. *Trends in Biotechnology*, 22, 455–462.

Graham, M. D. (1981). Efficiency comparison of 2 preparative mechanisms for magnetic separation of erythrocytes whole-blood. *Journal of Applied Physics*, 52, 2578–2580.

Haik, Y., Pai, V., and Chen, C. J. (1999). Development of magnetic device for cell separation. *Journal of Magnetism and Magnetic Materials*, 194, 254–261.

Han, K. H. and Frazier, A. B. (2004). Continuous magnetophoretic separation of blood cells in microdevice format. *Journal of Applied Physics*, 96, 5797–5802.

Han, K. H. and Frazier, A. B. (2005). Diamagnetic capture mode magnetophoretic microseparator for blood cells. *Journal of Microelectromechanical Systems*, 14, 1422–1431.

Han, K. H. and Frazier, A. B. (2006). Paramagnetic capture mode magnetophoretic microseparator for high efficiency blood cell separations. *Lab on a Chip*, 6, 265–273.

Hatch, G. P. and Stelter, R. E. (2001). Magnetic design considerations for devices and particles used for biological high-gradient magnetic separation (HGMS) systems. *Journal of Magnetism and Magnetic Materials*, 225, 262–276.

Hoffmann, C. and Franzreb, M. (2004). A novel repulsive-mode high-gradient magnetic separator—11. Separation model. *IEEE Transactions on Magnetics*, 40, 462–468.

Huang, R., Barber, T. A., Schmidt, M. A., Tompkins, R. G., Toner, M., Bianchi, D. W., Kapur, R., and Flejter, W. L. (2008). A microfluidics approach for the isolation of nucleated red blood cells (NRBCS) from the peripheral blood of pregnant women. *Prenatal Diagnosis*, 28, 892–899.

Ichikawa, N., Katsuyama, Y., Nagasaki, Y., and Ichiki, T. (2005). Microfluidic devices integrated with permalloy micropatterns for bead-based assay. In Laurell, T., Nilsson, J., Jensen, K., and Harrison, D. J. (Eds.) *Micro Total Analysis Systems 2004*, Vol. 2, 384–386.

Inglis, D. W., Riehn, R., Austin, R. H., and Sturm, J. C. (2004). Continuous microfluidic immunomagnetic cell separation. *Applied Physics Letters*, 85, 5093–5095.

Inglis, D. W., Riehn, R., Sturm, J. C., and Austin, R. H. (2006). Microfluidic high gradient magnetic cell separation. *Journal of Applied Physics*, 99, 08K101.

Jones, T. B. (1995). *Electromechanics of Particles*. Cambridge, UK: Cambridge University Press.

Jung, J. and Han, K. H. (2008). Lateral-driven continuous magnetophoretic separation of blood cells. *Applied Physics Letters*, 93.

Kim, K. S. and Park, J. K. (2005). Magnetic force-based multiplexed immunoassay using superparamagnetic nanoparticles in microfluidic channel. *Lab on a Chip*, 5, 657–664.

Kim, Y., Hong, S., Lee, S. H., Lee, K., Yun, S., Kang, Y., Paek, K. K., Ju, B. K., and Kim, B. (2007). Novel platform for minimizing cell loss on separation process: Droplet-based magnetically activated cell separator. *Review of Scientific Instruments*, 78, 074301.

Lee, C. S., Lee, H., and Westervelt, R. M. (2001). Microelectromagnets for the control of magnetic nanoparticles. *Applied Physics Letters,* 79, 3308–3310.

Lee, H., Liu, Y., Alsberg, E., Ingber, D. E., Westervelt, R. M., and Ham, D. (2005). An IC/microfluidic hybrid microsystem for 2D magnetic manipulation of individual biological cells. *Digest of Technical Papers—IEEE International Solid-State Circuits Conference,* 80.

Lee, H., Purdon, A. M., and Westervelt, R. M. (2004a). Manipulation of biological cells using a microelectromagnet matrix. *Applied Physics Letters,* 85, 1063–1065.

Lee, H., Purdon, A. M., and Westervelt, R. M. (2004b). Micromanipulation of biological systems with microelectromagnets. *IEEE Transactions on Magnetics,* 40, 2991–2993.

Liakopoulos, T. M., Choi, J. W., and Ahn, C. H. (1997). A bio-magnetic bead separator on glass chips using semi-encapsulated spiral electromagnets. *Transducers 97—1997 International Conference on Solid-State Sensors and Actuators, Digest of Technical Papers, Vols. 1 and 2.* 485–488.

Liu, Y. J., Guo, S. S., Zhang, Z. L., Huang, W. H., Baigl, D., Xie, M., Chen, Y., and Pang, D. W. (2007). A micropillar-integrated smart microfluidic device for specific capture and sorting of cells. *Electrophoresis,* 28, 4713–4722.

Lund-Olesen, T., Bruus, H., and Hansen, M. (2006). Passive magnetic separator integrated with microfluidic mixer: Demonstration of enhanced capture efficiency. *MEMS 2006: 19th IEEE International Conference on Micro Electro Mechanical Systems, Technical Digest,* 386–389.

Lund-Olesen, T., Bruus, H., and Hansen, M. F. (2007). Quantitative characterization of magnetic separators: Comparison of systems with and without integrated microfluidic mixers. *Biomedical Microdevices,* 9, 195–205.

Megens, M. and Prins, M. (2005). Magnetic biochips: a new option for sensitive diagnostics. *Journal of Magnetism and Magnetic Materials,* 293, 702–708.

Melville, D., Paul, F., and Roath, S. (1975). Direct magnetic separation of red-cells from whole-blood. *Nature,* 255, 706–706.

Melville, D., Paul, F., and Roath, S. (1982). Fractionation of blood components using high-gradient magnetic separation. *IEEE Transactions on Magnetics,* 18, 1680–1685.

Mirowski, E., Moreland, J., Russek, S. E., and Donahue, M. J. (2004). Integrated microfluidic isolation platform for magnetic particle manipulation in biological systems. *Applied Physics Letters,* 84, 1786–1788.

Mirowski, E., Moreland, J., Zhang, A., Russek, S. E., and Donahue, M. J. (2005). Manipulation and sorting of magnetic particles by a magnetic force microscope on a microfluidic magnetic trap platform. *Applied Physics Letters,* 86, 243901.

Moore, L. R., Zborowski, M., Nakamura, M., McCloskey, K., Gura, S., Zuberi, M., Margel, S., and Chalmers, J. J. (2000). The use of magnetite-doped polymeric microspheres in calibrating cell tracking velocimetry. *Journal of Biochemical and Biophysical Methods,* 44, 115–130.

Moreland, J., Mirowski, E., and Russek, S. E. (2005). Microfabricated spin-valve traps for manipulation of individual magnetic beads in a microfluidic environment. *Nanomedicine,* 1, 280.

Nakamura, M., Decker, K., Chosy, J., Comella, K., Melnik, K., Moore, L., Lasky, L. C., Zborowski, M., and Chalmers, J. J. (2001a). Separation of a breast cancer cell line from human blood using a quadrupole magnetic flow sorter. *Biotechnology Progress,* 17, 1145–1155.

Nakamura, M., Zborowski, M., Lasky, L. C., Margel, S., and Chalmers, J. J. (2001b). Theoretical and experimental analysis of the accuracy and reproducibility of cell tracking velocimetry. *Experiments in Fluids,* 30, 371–380.

Nandy, K., Chaudhuri, S., Ganguly, R., and Puri, I. K. (2008). Analytical model for the magnetophoretic capture of magnetic microspheres in microfluidic devices. *Journal of Magnetism and Magnetic Materials,* 320, 1398–1405.

Nedelcu, S. and Watson, J. H. P. (2002). Magnetic separator with transversally magnetised disk permanent magnets. *Minerals Engineering,* 15, 355–359.

Oberteuffer, J. A. (1973). High gradient magnetic separation. *IEEE Transactions on Magnetics,* MAG-9, 303–306.

Ostergaard, S., Blankenstein, G., Dirac, H., and Leistiko, O. (1999). A novel approach to the automation of clinical chemistry by controlled manipulation of magnetic particles. *Journal of Magnetism and Magnetic Materials,* 194, 156–162.

Pamme, N., Eijkel, J. C. T., and Manz, A. (2006). On-chip free-flow magnetophoresis: Separation and detection of mixtures of magnetic particles in continuous flow. *Journal of Magnetism and Magnetic Materials,* 307, 237–244.

Pamme, N. and Manz, A. (2004). On-chip free-flow magnetophoresis: Continuous flow separation of magnetic particles and agglomerates. *Analytical Chemistry,* 76, 7250–7256.

Pamme, N. and Wilhelm, C. (2006). Continuous sorting of magnetic cells via on-chip free-flow magnetophoresis. *Lab on a Chip,* 6, 974–980.

Pankhurst, Q. A., Connolly, J., Jones, S. K., and Dobson, J. (2003). Applications of magnetic nanoparticles in biomedicine. *Journal of Physics D—Applied Physics,* 36, R167–R181.

Paul, F., Melville, D., Roath, S., and Warhurst, D. C. (1981a). A bench top magnetic separator for malarial parasite concentration. *IEEE Transactions on Magnetics,* 17, 2822–2824.

Paul, F., Roath, S., Melville, D., Warhurst, D. C., and Osisanya, J. O. S. (1981b). Separation of malaria-infected erythrocytes from whole-blood-Use of a selective high-gradient magnetic separation technique. *Lancet,* 2, 70–71.

Pekas, N., Granger, M., Tondra, M., Popple, A., and Porter, M. D. (2005). Magnetic particle diverter in an integrated microfluidic format. *Journal of Magnetism and Magnetic Materials,* 293, 584–588.

Powles, R., Mehta, J., Kulkarni, S., Treleaven, J., Millar, B., Marsden, J., Shepherd, V., Rowland, A., Sirohi, B., Tait, D., Horton, C., Long, S., and Singhal, S. (2000). Allogeneic blood and bone-marrow stem-cell transplantation in haematological malignant diseases: A randomised trial. *Lancet,* 355, 1231–1237.

Qu, B. Y., Wu, Z. Y., Fang, F., Bai, Z. M., Yang, D. Z., and Xu, S. K. (2008). A glass microfluidic chip for continuous blood cell sorting by a magnetic gradient without labeling. *Analytical and Bioanalytical Chemistry,* 392, 1317–1324.

Ramadan, Q., Samper, V., Poenar, D., and Yu, C. (2004). On-chip micro-electromagnets for magnetic-based bio-molecules separation. *Journal of Magnetism and Magnetic Materials,* 281, 150–172.

Ramadan, Q., Samper, V., Poenar, D., and Yu, C. (2006a). Magnetic-based microfluidic platform for biomolecular separation. *Biomedical Microdevices,* 8, 151–158.

Ramadan, Q., Samper, V., Poenar, D. P., and Yu, C. (2006b). An integrated microfluidic platform for magnetic microbeads separation and confinement. *Biosensors and Bioelectronics,* 21, 1693–1702.

Ramadan, Q., Yu, C., Samper, V., and Poenar, D. P. (2006c). Microcoils for transport of magnetic beads. *Applied Physics Letters,* 88, 032501.

Rida, A., Fernandez, V., and Gijs, M. A. M. (2003). Long-range transport of magnetic microbeads using simple planar coils placed in a uniform magnetostatic field. *Applied Physics Letters,* 83, 2396–2398.

Rong, R., Choi, J. W., and Ahn, C. H. (2003). A functional magnetic bead/biocell sorter using fully integrated magnetic micro/nano tips. *MEMS-03: IEEE the Sixteenth Annual International Conference on Micro Electro Mechanical Systems.*

Safarik, I. and Safarikova, M. (1999). Use of magnetic techniques for the isolation of cells. *Journal of Chromatography B,* 722, 33–53.

Safarik, I. and Safarikova, M. (2004). Magnetic techniques for the isolation and purification of proteins and peptides. *BioMagnetic Research and Technology,* 2, 7.

Setchell, C. H. (1985). Magnetic separations in biotechnology—A review. *Journal of Chemical Technology and Biotechnology B—Biotechnology,* 35, 175–182.

Siegel, A. C., Shevkoplyas, S. S., Weibel, D. B., Bruzewicz, D. A., Martinez, A. W., and Whitesides, G. M. (2006). Cofabrication of electromagnets and microfluidic systems in poly(dimethylsiloxane). *Angewandte Chemie—International Edition,* 45, 6877–6882.

Smistrup, K. (2007). *Magnetic Separation in Microfluidic Systems.* Lyngby, Technical University of Denmark.

Smistrup, K., Bruus, H., and Hansen, M. F. (2007). Towards a programmable magnetic bead microarray in a microfluidic channel. *Journal of Magnetism and Magnetic Materials,* 311, 409–415.

Smistrup, K., Bu, M. Q., Wolff, A., Bruus, H., and Hansen, M. F. (2008). Theoretical analysis of a new, efficient microfluidic magnetic bead separator based on magnetic structures on multiple length scales. *Microfluidics and Nanofluidics,* 4, 565–573.

Smistrup, K., Hansen, O., Bruus, H., and Hansen, M. F. (2005a). Magnetic separation in microfluidic systems using microfabricated electromagnets-experiments and simulations. *Journal of Magnetism and Magnetic Materials,* 293, 597–604.

Smistrup, K., Hansen, O., Tang, P. T., and Hansen, M. F. (2005b). Selective magnetic bead capture using an addressable on-chip electromagnet array. In Laurell, T., Nilsson, J., Jensen, K., Harrison, D. J., and Kutter, J. P. (Eds.) *Micro Total Analysis Systems 2004,* Vol. 1, 509–511.

Smistrup, K., Kjeldsen, B. G., Reimers, J. L., Dufva, M., Petersen, J., and Hansen, M. F. (2005c). On-chip magnetic bead microarray using hydrodynamic focusing in a passive magnetic separator. *Lab on a Chip,* 5, 1315–1319.

Smistrup, K., Lund-Olesen, T., Hansen, M. F., and Tang, P. T. (2006a). Microfluidic magnetic separator using an array of soft magnetic elements. *Journal of Applied Physics,* 99.

Smistrup, K., Tang, P. T., Hansen, O., and Hansen, M. F. (2006b). Micro electromagnet for magnetic manipulation in lab-on-a-chip systems. *Journal of Magnetism and Magnetic Materials,* 300, 418–426, 08P102.

Spees, W. M., Yablonskiy, D. A., Oswood, M. C., and Ackerman, J. J. H. (2001). Water proton MR properties of human blood at 1.5 Tesla: Magnetic susceptibility, T-1, T-2, T-2* and non-Lorentzian signal behavior. *Magnetic Resonance in Medicine,* 45, 533–542.

Sugar, I. P. and Neumann, E. (1984). Stochastic model for electric field-induced membrane pores electroporation. *Biophysical Chemistry,* 19, 211–225.

Sung, H. W. F. and Rudowicz, C. (2003). Physics behind the magnetic hysteresis loop—a survey of misconceptions in magnetism literature. *Journal of Magnetism and Magnetic Materials,* 260, 250–260.

Takayasu, M., Kelland, D. R., and Minervini, J. V. (2000). Continuous magnetic separation of blood components from whole blood. *IEEE Transactions on Applied Superconductivity,* 10, 927–930.

Tamanaha, C. R., Mulvaney, S. P., Rife, J. C., and Whitman, L. J. (2008). Magnetic labeling, detection, and system integration. *Biosensors and Bioelectronics,* 24, 1–13.

Todd, P., Cooper, R. P., Doyle, J. F., Dunn, S., Vellinger, J., and Deuser, M. S. (2001). Multistage magnetic particle separator. *Journal of Magnetism and Magnetic Materials,* 225, 294–300.

Tondra, M., Granger, M., Fuerst, R., Porter, M., Nordman, C., Taylor, J., and Akou, S. (2001). Design of integrated microfluidic device for sorting magnetic beads in biological assays. *IEEE Transactions on Magnetics,* 37, 2621–2623.

Watson, J. H. P. (1973). Magnetic filtration. *Journal of Applied Physics,* 44, 4209–4213.

Weissman, I. L. (2000). Translating stem and progenitor cell biology to the clinic: Barriers and opportunities. *Science,* 287, 1442–1446.

Williams, P. S., Zborowski, M., and Chalmers, J. J. (1999). Flow rate optimization for the quadrupole magnetic cell sorter. *Analytical Chemistry,* 71, 3799–3807.

Wirix-Speetjens, R. and De Boeck, J. (2004). On-chip magnetic particle transport by alternating magnetic field gradients. *IEEE Transactions on Magnetics,* 40, 1944–1946.

Wirix-Speetjens, R., Fyen, W., De Boeck, J., and Borghs, G. (2006). Enhanced magnetic particle transport by integration of a magnetic flux guide: Experimental verification of simulated behavior. *Journal of Applied Physics,* 99.

Wirix-Speetjens, R., Fyen, W., Kaidong, X., Jo De, B., and Borghs, G. (2005). A force study of on-chip magnetic particle transport based on tapered conductors. *Magnetics, IEEE Transactions on,* 41, 4128–4133.

Xia, N., Hunt, T., Mayers, B., Alsberg, E., Whitesides, G., Westervelt, R., and Ingber, D. (2006). Combined microfluidic-micromagnetic separation of living cells in continuous flow. *Biomedical Microdevices,* 8, 299–308.

Yang, L. Y., Lang, J. C., Balasubramanian, P., Jatana, K. R., Schuller, D., Agrawal, A., Zborowski, M., and Chalmers, J. J. (2009). Optimization of an enrichment process for circulating tumor cells from the blood of head and neck cancer patients through depletion of normal cells. *Biotechnology and Bioengineering,* 102, 521–534.

Zborowski, M., Ostera, G. R., Moore, L. R., Milliron, S., Chalmers, J. J., and Schechter, A. N. (2003). Red blood cell magnetophoresis. *Biophysical Journal,* 84, 2638–2645.

Zborowski, M., Sun, L. P., Moore, L. R., Williams, P. S., and Chalmers, J. J. (1999). Continuous cell separation using novel magnetic quadrupole flow sorter. *Journal of Magnetism and Magnetic Materials,* 194, 224–230.

Zborowski, M., Tada, Y., Malchesky, P. S., and Hall, G. S. (1993a). Dark-field microscopy analysis of the magnetic deposition of bacteria on a glass-surface. *Colloids and Surfaces A—Physicochemical and Engineering Aspects,* 77, 209–218.

Zborowski, M., Tada, Y., Malchesky, P. S., and Hall, G. S. (1993b). Quantitative and qualitative analysis of bacteria in Er(III) solution by thin-film magnetophoresis. *Applied and Environmental Microbiology,* 59, 1187–1193.

Zhang, H. D., Moore, L. R., Zborowski, M., Williams, P. S., Margel, S., and Chalmers, J. J. (2005). Establishment and implications of a characterization method for magnetic nanoparticle using cell tracking velocimetry and magnetic susceptibility modified solutions. *Analyst,* 130, 514–527.

8 Microfluidic Manipulation of Biological Samples

Fridolin Okkels
Department of Micro- and Nanotechnology,
Technical University of Denmark, Lyngby

CONTENTS

For many years, researchers have manipulated biological samples on the chemical level, by observing responses on adding different kinds of signal molecules (e.g., hormones and cytokines). In recent years, this field has flourished dramatically with the aid of microfluidic systems, that is, integrating submillimeter channels into systems, which control and manipulate fluid samples of volumes down to nanoliter scale. Not only can the sample volumes be reduced dramatically, but the dosing can be targeted with subcellular resolution. Different components in a sample can be sorted out prior or after analysis—with the whole process occurring on a single chip.

This chapter describes the main building blocks within microfluidic manipulation of biological samples that will form the backbone of many microdevices in the years to come.

8.1   LAMINAR FLOWS

A prominent and important property of small-scale fluid flows is the smooth and undisturbed motion that characterizes laminar flows. These properties are advantageous for transport and handling of distinct portions of fluid, and when working with well-defined fluid interfaces, but complicate rapid mixing of small sample volumes, which for example, is essential for, characterizing fast reaction kinetics. This flow behavior should be seen in contrast to the spontaneous instabilities, eddies, and irregular flows, which are normally associated with larger-scale turbulent flows. Since practically all microfluidic flows are laminar, there is no need here to further address the transition from laminar to turbulent flows.[1]

8.2   STREAMLINES AND PARTICLE TRACES

Due to laminar flow, the trace of a minute particle released in a microfluidic flow will describe a nice well-defined curve, and for steady-state flows, having no time variations, this curve will be called a *streamline*.[2] Now, if the size and weight of the advected[3] particle increases, inertia will make it harder for the particle to follow any variation in the flow direction, and we now call the resulting new particle trajectory a *particle trace*. A nice everyday example illustrating the difference between these types of curves is the insects caught on the windshield of cars. While small light particles, for example, snowflakes, mostly follow the air current over the moving car, denser insects cannot follow the abrupt change in flow direction and collide onto the windshield. The difference between the snowflake and the insects is that they have different response times for how fast they adapt to changes in the flow direction. A shorter response time means a closer match between the particle trajectory and the streamlines.

In the case of biological samples in microfluidic flow, we estimate the characteristic relaxation time τ by looking at a neutrally buoyant particle of diameter $d_p = 5$ μm in water (density $\rho_p = 1\%10^3$ kg/m³, and dynamic viscosity $\eta = 1\%10^{-3}$ Pa s), where τ is then given by

$$\tau = \frac{d_p^2 \rho_p}{18\eta} = 1.5\mu s \qquad (8.1)$$

This shows that such a particle will always follow the surrounding fluid flow, as it is unlikely that it will experiences a change in flow direction within the timescale of 1.5 ηs in a microfluidic system.

8.3   STRUCTURE OF THIS CHAPTER

Microfluidic manipulation of biological samples can be performed by methods on all levels of complexity, and this chapter will begin at the simplest level by showing how microfluidic flows alone can deliver membrane permeable molecules with subcellular resolution. Methods of increasing complexity are then presented, such as ones incorporating wall interaction and diffusion-related effects, ending up with an example of creating monodisperse microdroplets for parametric study of protein crystallization. Furthermore, important and very useful theoretical concepts are introduced, both to aid the description of methods and to provide some tools for estimating basic system properties and responses.

Although some years old, one of the best reviews on actuation and manipulation in microfluidic systems is "Microfluidics: Fluid physics at the nanoliter scale" by Todd M. Squires and Stephen R. Quake from 2005 (Squires and Quake, 2005). This review has been a source of inspiration in choosing the different illustrative examples for this chapter.

8.4   PURE MICROFLUIDIC MANIPULATION

The following examples nicely illustrate that great scientific work may arise from simple microfluidic manipulation of some essential entities in biology: single cells and strings of DNA.

8.4.1   DIRECT UTILIZATION OF SIMPLE LAMINAR FLOWS

When manipulating cells, researchers have generally had the problem that bioactive reagents could only be added for a longer period to whole cells, while localization of reagents within parts of a cell has been hard to maintain due to rapid diffusion. By using light-triggered activation of photosensitive reagents, species gradients across single cells could be created (Adams and Tsien, 1993), but once the bioactive molecules was released in one subset of the cell, diffusion spread them within seconds (Takayama et al., 2003).

Takayama et al. (2001) overcame these difficulties by applying a simple and yet efficient technique, in which they created steady reagent gradients along a microchannel, by having two or more upstream inlets, which are joined side by side to create the main straight micro channel on which the cells are attached. The reagent enters from one of these inlets, while the remaining ones supply pure buffer, as seen in Figure 8.1. Because of the laminar flow, the different flows continue smoothly downstream through the microchannel, and as the cells are placed close to the junction, the interface between the different fluids is still distinct.

In a paper from 2003, Takayama et al. (2003) exploit this technique further by investigating different kinds of phenomena on the subcellular level. The reported investigations cover, for example, the local dynamics of intercellular components (mitochondria) using fluorophores, receptor-mediated endocytosis, and detachment of a BCE cell after digestion of some of its underlying extracellular matrix protein with trypsin, as showed in Figure 8.2.

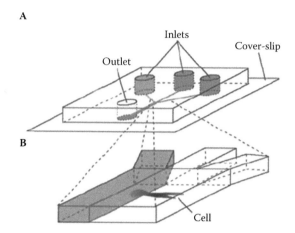

FIGURE 8.1 Schematic illustration of the channel layout and the technique for creating subcellular positioning of bioactive reagents. The reagent (gray) enters from one of the inlets, while the remaining inlets supply pure buffer, and not far downstream from the inlet junction, a strong persistent reagent gradient is created due to the laminar flow and the limited diffusive spread. (Reprinted from Takayama, S., Ostuni, E., Leduc, P., Naruse, K., Ingber, D. E., and Whitesides, G. M. *Nature,* 411, 1016–1016, 2001. Macmillan Publishers Ltd.: Copyright 2001. With permission.)

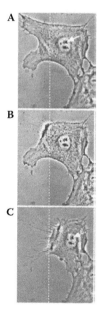

FIGURE 8.2 Detachment of a BCE cell after adding trypsin to the flow left of the interface (dashed line). (A) Before treatment, (B) after 1 min of treatment, and (C) after 3.5 min of treatment. (Reprinted from Takayama, S., Ostuni, E., Leduc, P., Naruse, K., Ingber, D. E., and Whitesides, G. M. *Chemistry and Biology,* 10, 123–130, 2003. Macmillan Publishers Ltd: Copyright 2003. With permission.)

8.4.2 DIFFERENT WAYS OF STRETCHING DNA

Another example of pure microfluidic manipulation is the stretching of DNA directly by the action of an elongational flow field, where the DNA has no direct contact with the channel walls as they are only used for creating the specific flow field. Such stretching or unfolding of polymer chains in elongational flow was investigated theoretically by Degennes in 1974 (Degennes, 1974), predicting an unfolding transition for high shearing flows, involving strong velocity gradients. Early observations of such polymer unfolding in elongational flows were complicated and indirect, for example, by use of light scattering (Menasveta and Hoagland, 1991), but by using fluorescently labeled DNA molecules, Perkins and Degennes (Degennes, 1974, Perkins et al., 1997) observed the unfolding of individual chains in the high shearing flow field. Furthermore, they could directly visualize and characterize how the initial folding of the chains altered the speed at which they unfolded, as seen in Figure 8.3.

A completely different way of stretching out DNA will be mentioned here, even though it uses direct channel confinement to unfold the long chains, and is thus diametrically different from pure microfluidic manipulation. The method was applied by J. O. Tegenfeldt in 2004 (Tegenfeldt et al., 2004) where DNA was led into nanometer-sized channels, which prevented the polymer to pass in any entangled form. Here, the flow field only acts to force the DNA through the channels. Once again, theoretical work by Degennes (1979) was essential for the interpretation of the measurements, since he demonstrated that the extension of the polymer in the channel must be proportional to the contour length[4] of the chain (Figure 8.4).

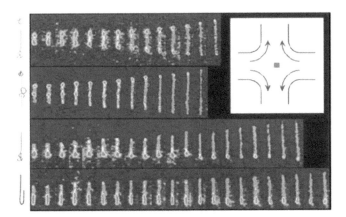

FIGURE 8.3 (See color insert) Stretching of DNA by an elongational flow, visualized by using fluorescently labeled DNA molecules. Each row of images show the unfolding of different molecular configurations in time steps of 0.13 s. The four examples form the classification of configurations, depicted in from of every row and denoted from top to bottom: dumbbell, kinked, half-dumbbell, and folded. (Inset) The channel layout used to create the elongational flow, with the observation region in the middle. (Perkins, T. T., Smith, D. E., and Chu, S. *Science,* 276, 2016–2021, 1997. With permission.)

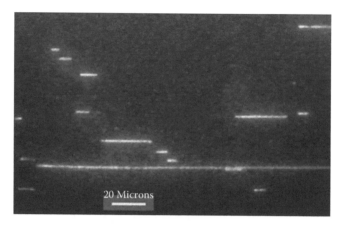

FIGURE 8.4   Stretching of fluorescently labeled DNA by confining it in narrow nanochannels. Only the DNA chains are visible in the picture, while the horizontal nanochannels or the convective fluid are not seen. (Reprinted from Tegenfeldt, J. O., Prinz, C., Cao, H., Chou, S., Reisner, W. W., Riehn, R., Wang, Y. M., Cox, E. C., Sturm, J. C., Silberzan, P., and Austin, R. H. *Proceedings of the National Academy of Sciences of the United States of America,* 101, 10979–10983, 2004. Copyright 2004. With permission.)

8.5   COMBINED EFFECT OF STREAMLINES AND DIRECT WALL INTERACTION

Until now, we have covered devices or techniques in which the role of microfluidic channels have been either to create specific flows, which are used in the manipulation process, or to directly confine and thereby manipulate coiled polymer chains.

In the following, we present two particle manipulation methods based on the combined action from specific flows and direct wall interaction, where both actions are necessary. These methods utilize the effect that particles in bulk flow follow their center of particle streamline,[5] while channel walls or other fixed obstacles may influence which streamline the particle follows, depending on the physical size of the particle.

8.5.1   PINCHED FLOW FRACTIONATION (PFF)

Many techniques exist for particle separation, which can also be applied for separation of macromolecules (ranging from 1 nm to 100 µm in diameter), but the different techniques have their limitations, such as the requirement of outer fields, for example, gravitational, centrifugal, electrical, thermal, or cross-flow field,[6] or the restriction of only using distinct samples, such that the separation process is not continuous.[7]

In 2004, M. Yamada, M. Nakashima, and M. Seki came up with a very simple and elegant fractionation technique called *pinched flow fractionation* (PFF) (Yamada et al., 2004), which is both continuous and without the requirement of outer fields. In PFF, the different particles to be separated are positioned to flow very close to one of the channel walls and, by this confinement, their center of the particle follows different streamlines, which then lead to a subsequent separation.

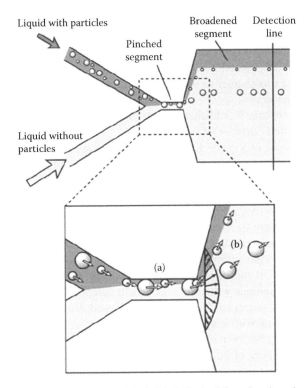

FIGURE 8.5 Illustration of the principle behind pinched flow fractionation, where the top part shows the complete device, while the lower part focuses on the separation process in the pinched segment. (Reprinted from Yamada, M., Nakashima, M., and Seki, M. *Analytical Chemistry,* 76, 5465–5471, 2004. Copyright 2004. American Chemical Society. With permission.)

The realization of this principle is showed in Figure 8.5. The top part shows the complete device, where two liquids, with and without particles, enters the pinched channel segment from each inlet. Then, by controlling the ratio of inlet flow rates, the particles are focused on one sidewall, as showed in the lower enlargement, such that their center of particle necessarily have to be at different distances from the sidewall, depending on their size. Then as the channel is broadened after the pinched segment, each particle freely follows its corresponding streamline, and the new distance to the sidewall of the broadened segment is roughly amplified by the ratio between the widths of the pinched to the broadened channels. Yamada et al. show also that the separation is unaffected by variations in the total flow rate, as long as the ratio of the two inlet flow rates is kept constant.

In the same work by Yamada, they also extend the functionality of the PFF by turning it into a sorting device. This was done by making a number of branches at the end of the broadened segment that all lead to specific outlets, such that the separated particles could be collected independently at the different outlets. Still, recent studies conclude that implementing PFF for particle sorting requires well-defined and steady-flow conditions, and this may be hard to accomplish when used as part of larger lab-on-a-chip system.[8]

This step of turning a microfluidic separating method into a sorting device is not limited to the case of pinched flow fractionation, but is generally used, and seen later in Section 8.7.1 of this chapter as the separation method of T-sensors are used in the H-filter sorting device.

Once a new technique is published, it is natural to progress by refining it, and while Yamada et al. sorted particles of size down to 15 µm, particles of size 0.25 µm was separated in the work of Vig and Kristensen (2008). They furthermore presented and demonstrated an enhancement of the method by spitting up the broadened segment into two, by a snakelike structure, as shown in Figure 8.6. This enhanced pinched flow fractionation (EPFF) design works by simply altering the fluid flow in the broadened segment and thereby improving the separation at no expense. The improvement depends on the particle size, and is measured to be in the range of 40%–75%.

Compared to other branches of fluid mechanics, microfluidics are well suited for theoretical studies as the laminar flow conditions and the simple channel geometries make it possible to find analytic solutions to the governing equations of fluid motion. These governing equations will not be dealt with here, but an excellent introduction to theoretical microfluidics has been written by Henrik Bruus (2008).

Many have come up with simple estimates of the sorting properties of PFF devices (Takagi et al., 2005), but even though such considerations may provide approximate answers, the investment of building theoretical models of the sorting mechanism, based on solutions to the governing equations of fluid motions, is indeed rewarding as they do a much better job (Andersen et al., 2009).

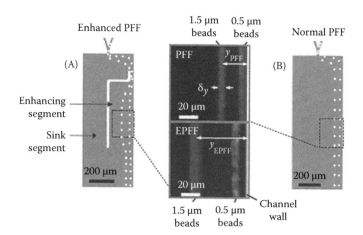

FIGURE 8.6 Illustration of the enhanced pinched flow fractionation (EPFF) device (A) in correspondence to the original pinched flow fractionation (PFF) device (B), with picture of the real experiment shown in the center. (Reprinted from Vig, A. L. and Kristensen, A. *Applied Physics Letters*, 93, 3, 2008. Copyright 2008, American Institute of Physics. With permission.)

8.5.2 DETERMINISTIC LATERAL DISPLACEMENT DEVICES (MICROFLUIDIC BUMPER ARRAYS)

Another aspect of particle separation is the resolution at which different particles can be separated, that is, the variation of the particle sizes within one category of sorted particles. Earlier techniques for particle separation was, for example, exclusion (Dupont and Mortha, 2004), where the sample flows through large porous beads, such that smaller particles lacks behind due to longer diffusive paths, but because of the stochastic packing of the beads, even monodisperse particles would be resolved quite poorly by this technique.

In 2004, L. R. Huang et al. (2004) presented a sorting technique in which all particles follow a path deterministically depending on the particle size, and this leads to an impressive resolution. The technique is denoted deterministic lateral displacement, but the devices are commonly called *bumper arrays*, and the principle is illustrated in Figure 8.7. The particle separation occurs due to particle interaction with an obstacle matrix, consisting of rows of identical pillars through which the solution flows (orange arrow in Figure 8.7A). Small particles follow their corresponding streamlines, and from the periodicity in the arrangement of pillar rows, these streamlines zigzag through the matrix without any lateral displacement, as shown in Figure 8.7B. All streamlines end up in their initial lateral position after passing three rows. Larger particles, on the other hand, are too large to continuously follow their initial streamlines, since the passage between the pillars force the large particles to change streamline, as seen in Figure 8.7C. As a result, larger particles follow a displacement path instead of the zigzag path, which causes the particles to be separated and subsequently sorted depending on their size.

The two kinds of paths (zigzag and displacement) are called *transport modes*, and were experimentally observed with the use of fluorescent polystyrene microspheres with diameters of 0.40 and 1.03 μm in aqueous buffer; see Figure 8.8.

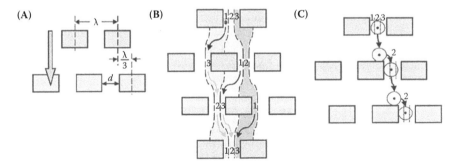

FIGURE 8.7 Principle of deterministic lateral displacement devices (so-called microfluidic bumper arrays). (A) Arrangement of solid pillars and overall flow direction ((A, B, and C) arrow). (B) Three characteristic fluid streams (horizontal) that pass through the upper gap, and recombines at a gap, three rows further downstream with similar transverse position. Small particles stay within their corresponding fluid streams, while larger particles, as seen in (C), are forced by the pillar arrangement to change fluid stream and thereby experience a lateral displacement. (Huang, L. R., Cox, E. C., Austin, R. H., and Sturm, J. C. *Science, 304,* 987–990, 2004. With permission.)

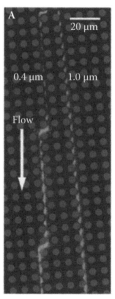

Zigzag Displacement

FIGURE 8.8   **(See color insert.)** Image of a deterministic lateral displacement device sorting fluorescent microspheres, and showing trajectories of the two transport modes. (Huang, L. R., Cox, E. C., Austin, R. H., and Sturm, J. C. *Science,* 304, 987–990, 2004. With permission.)

The high resolution of deterministic lateral displacement devices arise since the particles are continuously alignment during contact with the pillars, and this then prevents any stochastic noise from altering the particle paths. The overall periodicity of the pillar arrangement in the matrix then makes monodisperse particles follow exactly the same path, and as a result, the measured resolution of monodisperse particle separation is limited by the manufacturing inhomogeneities of the test particles.

Deterministic lateral displacement is now a well-established technique for particle separation, and one example of recent progress is L. Sasso et al., who in 2009 were able to obtain a separation of particles as small as 70 nm in radius (Sasso et al., 2009). When separating particles at these dimensions, Brownian motion cannot be neglected, and the effect of the resulting particle diffusion is dealt with theoretically in a paper by M. Heller and H. Bruus from 2008 (Heller and Bruus, 2008). They derive an extended model that helps correcting some systematic deviation between experimental measurements and the former model estimates.

8.6 DIFFUSION-DRIVEN MANIPULATION

All suspended particles at around room temperature experiences continuous bombardment of the surrounding fluid molecules and, while this normally does not influence the motion of suspended cells, some intracellular components are small enough that individual fluid molecule collisions can perturb the position, leading to Brownian motion. This motion will then result in a diffusive transport of the components with related diffusion coefficient D. As expected, the diffusion coefficient varies inversely proportional to the particle size, as characterized by the Einstein relation, originally published in 1905[9] (Einstein, 1905):

$$D = \frac{\kappa_B T}{6\pi a \eta} \qquad (8.2)$$

Here, a is the particle radius, T is the temperature in Kelvin, and η is the dynamic viscosity. The Boltzmann constant k_B relates particle energy with observed bulk temperature, such that $k_B T$ expresses the mean thermal energy per particle.

8.6.1 TYPES OF SPECIES TRANSPORT—THE PÉCLET NUMBER

Transport of molecular species in microfluidic channels comes mainly from two kinds of transport:

- *Advective transport*, where the molecules follow the flow of the surrounding fluid[10]. For a particle being advected with the flow speed U, through a microfluidic systems of dimension L, we can introduce the advective timescale $T_C = L/U$, which is nothing but the time it takes for the particle to "drift" through the system.
- *Diffusive transport*, can either consist of Brownian motion for medium-sized particles, or a diffusive flux, proportional to the concentration gradient (Fick's law) for substances described by a related concentration field. In both cases, a diffusive timescale is defined by $T_D = L^2/D$.

Many of the devices and techniques to be described in the following rely on the large differences in these two types of transport. Of the most prominent differences can be mentioned: Advective transport relates directly on the motion of the fluid and build up sharp changes in a concentration field. On the contrary, diffusive transport is flow independent[11] and smear out concentration variations.

With these profound differences in type of transport, their relative rate in a fluidic system gives valuable information about the general transport properties, and this leads directly to the Péclet number,[12] which is defined as the ratio of the advection rate to the diffusion rate. Since rates are inversely proportional to their related timescales, we can define the Péclet number in the following way:

$$Pe \equiv \frac{advective\ rate}{diffusive\ rate} = \frac{diffusion\ time}{advection\ time} = \frac{L^2/D}{L/U} = \frac{LU}{D} \qquad (8.3)$$

Here, the common length L to be used in both timescales is the transverse length scale, that is, in the direction of the diffusive flux.

Similarly to other dimensionless numbers such as the Reynolds number, the Péclet number (Pe) quantifies a general property of the system, and as a consequence, two

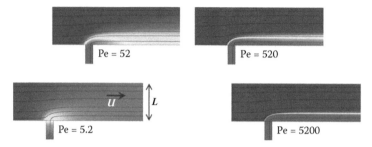

FIGURE 8.9 (See color insert.) Numerical example showing the influence of varying Péclet numbers. A side fluid inlet containing a diffusive substance is added to a main fluid flow, and the transition from diffusive transport to advective transport is clearly illustrated, as the Péclet number for the different examples increase in steps of one order of magnitudes. [Numerical simulations done in COMSOL by the author.]

different fluidic systems both fully characterized by the same Pe will be governed by the same type of species transport. For example, the species transport in systems with low Pe (Pe < 10) will dominantly be diffusive, with a marginal role played by the fluid flow, while high-Pe systems (Pe > 1000) rely heavily on the fluid flow and are characterized by sharp changes in concentration, where the fluids continue downstream as two separated flows. A graduate transition between the two types of transport occurs for intermediate Pe (10 > Pe > 1000).

Figure 8.9 shows the effect of varying Pe on a numerical model of how two miscible fluids merge downstream. A side fluid inlet containing a diffusive substance is added to a main fluid flow, and the transition from low-Pe diffusive transport to high-Pe advective transport is clearly illustrated.

Knowing the three characteristic quantities (L, U, and D) of a fluidic system, the Péclet number can be computed with minimal effort, and knowing its value gives direct information about the mixing properties of the system. This will then decide if diffusion does the job, or if additional means have to be taken to ensure good mixing. In that way, dimensionless quantities give quick and important estimates, valuable for all researchers within a large group of natural sciences.

8.7   DIFFUSION-DRIVEN DEVICES

Combining the well-controlled handling of fluids and fluid interfaces in laminar flows with diffusion of different species across such interfaces results in simple and yet effective devices, some of which will be presented in the following. Yet these same properties cause problems when quick mixing of substances is needed, and one ingenious, and now classic, way of overcoming this problem is presented.

8.7.1   T-SENSOR

A device that nicely illustrates the benefits of laminar flow conditions within microchannels is the T-sensor, which was originally investigated by Kamholz et al. (1999). In the T-sensor, streams of two different fluids join together into a straight shallow microchannel such that they meet in a single interface, extending downstream in the middle of the channel. Due to the laminar state, only diffusive mixing occurs between the two fluids, where the different molecules diffuse across the interface once the fluids come into contact. A schematic representation of the T-sensor is shown in Figure 8.10, where the merging fluids are marked by lighter grayscales. The region of the darkest grayscale (called the *interdiffusion region*) contains species from both fluids, and since the two fluids have different diffusion coefficients, the interdiffusion region is asymmetric. In the work of Kamholz, this region could be directly observed optically, as they merged two fluids containing the human serum albumin and the red fluorophore albumin blue 580, respectively, where albumin blue 580 has a particularly high affinity for serum albumins. In that way, they observed a clear increase in fluorescence intensity one the two species meet. From the simple geometry and flow characteristics of this device, it is possible to measure not only the diffusion coefficients of the different species but also the different quantities related to the reaction or binding kinetics between the two species, as exemplified by the work of Baroud et al. in 2003 (Baroud et al., 2003). These properties quickly made the T-sensor an important analysis tool (Weigl and Yager, 1999), and a few years later, it was also used with success in immunoassay analysis (Hatch et al., 2001).

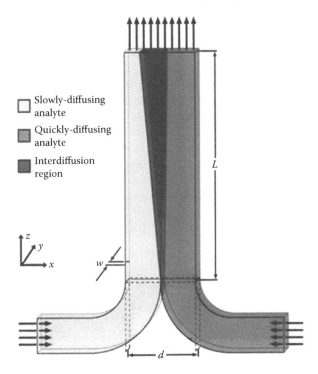

FIGURE 8.10   Illustration of the flow in the T-sensor with a steady-state flow with two input fluids, each containing one diffusive species. The interdiffusive region spread from the fluid interface and is shown in dark gray. (Reprinted from Kamholz, A. E., Weigl, B. H., Finlayson, B. A., and Yager, P. *Analytical Chemistry,* 71, 5340–5347, Copyright 1999. American Chemical Society. With permission.)

8.7.2 H-Filter

As mentioned earlier, a particle-separating device can be upgraded into a sorting device by splitting the particle-carrying flow into different outlets, and this is no exception with the T-sensor, whose related sorting device is called the H-filter. Contrary to the T-sensor, which is primarily used for studying chemical reaction, the H-filter sorts particles separated simply by diffusion. As illustrated schematically in Figure 8.11, two flows are joined together, where the upper one usually is a clear buffer, while the other contains particles of different diffusivity (smaller yellow and larger red particles), which are linked to differences in their size through the Einstein relation, introduced in Section 8.3. Similarly to the T-sensor, the particles will diffuse across the interface of the two fluids, and while larger particles (with low diffusivity) mainly stay in their initial fluid volume, smaller particles diffuse into the buffer flow in the combined channel. Once the two fluids flows are split further downstream, the most diffusive particles are now sorted from their heavier components. The simplicity of this technique may also point to the reason why the H-filter was reported first, three years before the T-sensor! The H-filter was originally mentioned among other microfluidic devices in a paper of Brody et al. (1996), and the use of device was extended further in a paper by Brody and Yager (Brody and Yager, 1997).

Once devices such as the ones presented earlier have been acknowledged by the scientific community, they are seldom used alone but contribute rather to a particular element in a larger process, and one nice example is a combination of an H-filter and a T-sensor presented in the paper by Schilling et al. (2002), and illustrated schematically in Figure 8.12. The process in this three-inlet, two-outlet device is first to join a flow of bacterial cells (*E. coli*) with a flow of lytic agent into the H-filter. The suspended cells lyse once they come into in contact with the lytic agent, and the released intracellular components then diffuse away from the cell stream and are sorted out by the H-filter as they flow toward the next step in the process. The presence of different intercellular components is then detected in the T-sensor by the creation of fluorescent products as detection molecules are added from the opposite inlet.

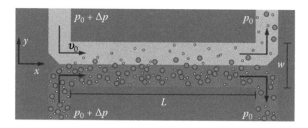

FIGURE 8.11 Schematic figure presenting the working principle of the H-filter. Large red particles diffuse slowly and stay thereby in the initial flow, while smaller yellow particles diffuse into the adjacent fluid flow from which they can be sorted out. (Reprinted from Bruus, H. *Theoretical Microfluidics* (Oxford Master Series in Condensed Matter Physics). Oxford, Oxford University Press, 2008. With permission.)

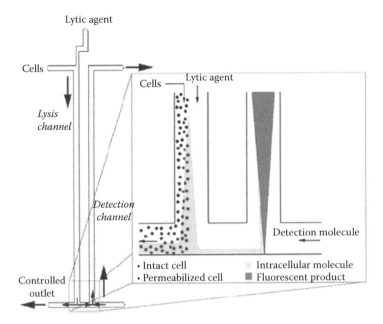

FIGURE 8.12 Schematic illustration of an H-filter attached to a T-sensor for detecting proteins extracted from lysed cells. (Reprinted from Schilling, E. A., Kamholz, A. E., and Yager, P. *Analytical Chemistry,* 74, 1798–1804, 2002. Copyright 2002. American Chemical Society. With permission.)

8.8 MIXING IN MICROFLUIDICS

The techniques and devices covered in this section are all based on diffusion as the main separating effect, and generally, diffusion plays an increasingly important role on smaller length scales. Still, one important diffusion-related fluid manipulation is causing problems at micron scales, and this is mixing, which is, for example, the ability of two different miscible fluids merged in the upstream part of a microchannel to reach a homogeneous distribution of both fluid substances further downstream.[13] This challenge is not caused by any reduction of the diffusive transport, but rather the absence of turbulent mixing in the laminar microflows. High Reynolds number turbulent flows, normally seen on macroscopic length scales, efficiently elongate and bend the fluid–fluid interfaces, such that diffusion across the different interfaces finally can create a homogeneous mixture of the two fluids. In simple laminar flows, diffusive transport only occurs across the straight interface, much similar to the case of the T-sensor, and this can result in substantial mixing times. For example, 30 base pair DNA molecules in water have a diffusion coefficient of $D \approx 4 \cdot 10^{-11}$ m^2/s, which will diffuse across a typical microfluidic distance ($L = 100$ µm) in the time $T_D = L^2/D = 250$ s ≈ 4 min.[14] Since the interface is along the flow, the dominant diffusion will be transverse to the channel and therefore will be largely unaffected by the simple flow.[15] A pure diffusive mixing of the DNA molecules in simple microchannels will therefore be achieved in timescales comparable to the diffusion time (i.e., 4 min). Such timescales are problematic to work with in otherwise-fast lab-on-a-chip systems. Especially, when seen in the light that simple geometrical modifications of the microchannels can reduce the mixing time to fractions of seconds, as described in the following.

8.8.1 MICROFLUIDIC MIXERS

The general aim in all mixing processes, whether on small or large length scales, is the following: *To reduce the distance the different substances have to diffuse through for a homogeneous state to arise.* In the case of turbulence, the instability and eddies do a good job in stretching and folding a fluid interface into thin layers, such that diffusion can finish the job. The situation is completely different in laminar microflows, as the low Reynolds number prevents any spontaneous instability from arising. Therefore, extra flow patterns have to be added to the main downstream flow to force the required stretching and folding of interfaces, and this is the basic ingredient in almost all microfluidic mixing devices.[16] These ideas were first presented and related to chaotic dynamical systems by Aref (1984).

Microfluidic mixing devices are categorized as being either active, where the additional flows are actuated by external forces or actions, or passive, where the shape of the channel itself makes the primary downstream flow give rise to the additional flows.

The direct need for mixing in microsystems made this research field a hot topic during the 1990s, where many types of both passive and active mixers were developed. This changed somehow in 2002 with the introduction of a simple, yet highly efficient passive micromixer by A. D. Stroock (Stroock et al., 2002), which was quickly named the *staggered herringbone mixer.* The groundwork for the mixer was laid in 2001 by the work of Ajderi (2002), where he investigated how transverse secondary flows could be produced by the downstream flow itself by patterning the floor of the channel with bas-relief structures. As an example, Figure 8.13 shows how a simple patterning of grooves on the channel floor creates a simple swirling flow, with a circular secondary flow.

The idea behind the staggered herringbone mixer is to produce a recurrent pattern of secondary flows in the channel, in such a way that fluid interfaces get stretched and folded, and the mixing thereby is greatly enhanced. Figure 8.14 shows the reoccurring pattern of grooves, and the two different kinds of secondary flows produced (Stroock and McGraw, 2004).

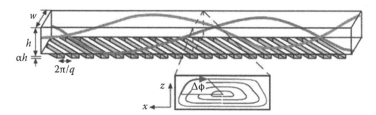

FIGURE 8.13 Schematic illustration of the secondary flows, generated by grooving the bottom of the channel. (Reprinted from Stroock, et al., Chaotic mixer far microchannels. *Science*, 295, 647–651 2002. With permission.)

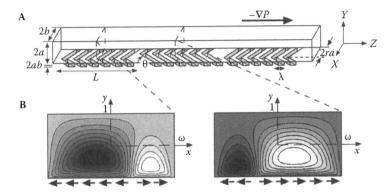

FIGURE 8.14 Design and working principle for the staggered herringbone mixer. By sequentially changing between two sections of different groove patterns, the different secondary flow patterns will deform the fluid interface. (Reprinted from Stroock, A. D. and McGraw, G. J., *Philosophical Transactions of the Royal Society of London A—Mathematical Physical and Engineering Sciences,* 362, 971–986, 2004. With permission.)

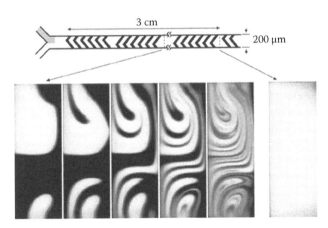

FIGURE 8.15 Effect of mixing at different stages downstream of the staggered herringbone mixer, measure by confocal microscopy. Last picture shows a well-mixed homogeneous state (Modified from Stroock, et al., *Science,* 295, 647–651. With permission from AAAS. 2002).

Figure 8.15 shows direct measurements by confocal microscopy of the mixing process, where initially one half of the channel was filled by a fluorescent solute. The concentration intensity in subsequent cross-section pictures nicely reveals the successively increased layering of the two fluids due to repeated stretching and folding, a process also denoted the "baker transformation." In the end, a nice homogeneous mixed state is achieved, without the use of any external actuation, and within a fraction of channel length used otherwise if the mixing had to rely on diffusion alone.

The efficiency of the staggered herringbone mixer was quickly recognized in the microfluidics communities worldwide, and it has now become an icon of the progress in the field of microfluidics. This mixer moderated the direct need for further development of passive micromixers, while the development of active mixers are still ongoing, since these mixers, for example, are still needed when microsamples needs to be mixed in localized containers and/or at specific time intervals.

8.9 SURFACE EFFECTS USED FOR MICROFLUIDIC MANIPULATION

It follows directly from dimensional arguments that, when going toward smaller length scales, surface-related effects, for example, diffusion, thermal conduction, adhesion, and surface tension, become increasingly more dominant compared to volume-related effects, for example, inertia, weight, and heat capacity (Wautelet, 2001). As a result, the utilization of surface effects in microfluidic systems results in very robust devices that are marginally affected by external influence (e.g., gravitation).

A nice example of such a microfluidic system was first studied by Anna et al. (2003), where they utilize a flow-focusing device to form monodisperse bubbles created by capillary instability. Figure 8.16 shows a picture of the device before operation, where the inlets of oil and water together with important length scales are marked. It is during the passage through the narrow central slit that the two oil–water interfaces become unstable and break up in drops.

The drop formation of the device is controlled by the oil flow rate Q_o and the ratio between the internal water flow rate to the external oil flow rate Q_i/Q_o, and a phase diagram for the drop formation is shown in Figure 8.17. Each picture shows the formation related to the specified value of Q_o (rows) and Q_i/Q_o (columns), and the phase diagram shows a rich variety of drop formation dynamics, which in some cases produce monodisperse drops.

This technique was quickly adapted by the analytic chemistry community to create small containers for microsample preparation and analysis. The method was, for example, modified to create small inclusions, which made it possible to measure millisecond kinetics by achieving a fast efficient mixing inside each inclusion drop (Song and Ismagilov, 2003). In another example, parametric study of protein

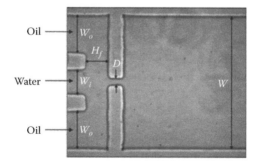

FIGURE 8.16 Picture of the flow focusing device before operation with annotation of the different inlets and geometrical distances. (Reprinted from Anna, S. L., Bontoux, N., and Stone, H. A. *Applied Physics Letters,* 82, 364–366, 2003. Copyright 2003, American Institute of Physics. With permission.)

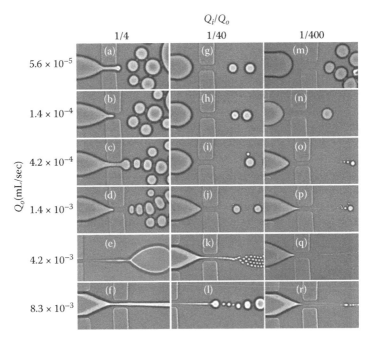

FIGURE 8.17 Phase diagram for drop formation by means of flow focusing. Each picture shows the formation related to the specified value of Q_o (rows) and Q_i/Q_o (columns). (Reprinted from Anna, S. L., Bontoux, N., and Stone, H. A. *Applied Physics Letters,* 82, 364–366, 2003. Copyright 2003, American Institute of Physics. With permission.)

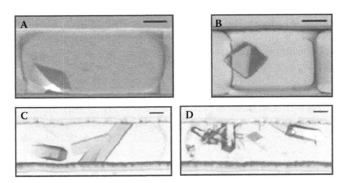

FIGURE 8.18 (See color insert.) Polarized light micrographs of protein crystals obtained inside droplets on a microfluidic chip. Scale 50 μm. Reprinted from Zheng, B., Roach, L. S., and Ismagilov, R. F. *Journal of the American Chemical Society,* 125, 11170–11171, 2003. Copyright 2003, American Chemical Society. With permission.)

crystallization inside small sample "inclusions" was conducted (Zheng et al., 2003; Zheng et al., 2005), where polarized light micrographs of different solute composition are shown in Figure 8.18. In this way, massive amounts of screenings for combinatorial chemistry can be prepared and analyzed with sample sizes measured in nanoliters.

8.10 FINAL REMARKS ON MICROFLUIDIC MANIPULATION OF BIOLOGICAL SAMPLES

The topics and techniques covered in this chapter form the basic building blocks of how to manipulate biological samples using different combinations of confined microfluidic flows, diffusion, and simple surface effects. As mentioned in the first part of the chapter, this field is nicely covered by the review of Todd M. Squires and Stephen R. Quake from 2005, and the further advances in this subject in the following five years are addressed in a critical review by Marre and Jensen (2010). This critical review deals mainly with a higher-level combination of the different phenomena and techniques described in the 2005 review, and this may be an indicator of the present direction in microfluidic systems research. Once a full and comprehensive toolbox is built up, it is natural to combine them into systems of increasing complexity, and thus the coming years will likely give us microfluidic systems that finally and literally will live up to the notion "lab-on-a-chip."

REFERENCES

Adams, S. R. and Tsien, R. Y. (1993). Controlling cell chemistry with caged compounds. *Annual Review of Physiology,* 55, 755–784.

Ajdari, A. (2002). Transverse electrokinetic and microfluidic effects in micropatterned channels: Lubrication analysis for slab geometries. *Physical Review E,* 65, 9.

Andersen, K. B., Levinsen, S., Svendsen, W. E., and Okkels, F. (2009). A generalized theoretical model for "continuous particle separation in a microchannel having asymmetrically arranged multiple branches." *Lab on a Chip,* 9, 1638–1639.

Anna, S. L., Bontoux, N., and Stone, H. A. (2003). Formation of dispersions using "flow focusing" in microchannels. *Applied Physics Letters,* 82, 364–366.

Aref, H. (1984). Stirring by chaotic advection. *Journal of Fluid Mechanics,* 143, 1–21.

Baroud, C. N., Okkels, F., Menetrier, L., and Tabeling, P. (2003). Reaction-diffusion dynamics: Confrontation between theory and experiment in a microfluidic reactor. *Physical Review E,* 67, 4.

Brody, J. P. and Yager, P. (1997). Diffusion-based extraction in a microfabricated device. *Sensors and Actuators A-Physical,* 58, 13–18.

Brody, J. P., Yager, P., Goldstein, R. E., and Austin, R. H. (1996). Biotechnology at low Reynolds numbers. *Biophysical Journal,* 71, 3430–3441.

Bruus, H. (2008). *Theoretical Microfluidics* (Oxford Master Series in Condensed Matter Physics). Oxford, Oxford University Press.

Degennes, P. G. (1979). *Scaling Concepts in Polymer Physics,* New York, Cornell University Press.

Degennes, P. G. (1974). Coil-stretch transition of dilute flexible polymers under ultrahigh velocity-gradients. *Journal of Chemical Physics,* 60, 5030–5042.

Dupont, A. L. and Mortha, G. (2004). Comparative evaluation of size-exclusion chromatography and viscometry for the characterisation of cellulose. *Journal of Chromatography A,* 1026, 129–141.

Einstein, A. (1905). Über die von der molekularkinetischen Theorie der Wärme geforderte Bewegung von in ruhenden Flüssigkeiten suspendierten Teilchen. *Annalen der Physik,* 17, 549–560.

Hatch, A., Kamholz, A. E., Hawkins, K. R., Munson, M. S., Schilling, E. A., Weigl, B. H., and Yager, P. (2001). A rapid diffusion immunoassay in a T-sensor. *Nature Biotechnology,* 19, 461–465.

Heller, M. and Bruus, H. (2008). A theoretical analysis of the resolution due to diffusion and size dispersion of particles in deterministic lateral displacement devices. *Journal of Micromechanics and Microengineering*, 18, 6.

Huang, L. R., Cox, E. C., Austin, R. H., and Sturm, J. C. (2004). Continuous particle separation through deterministic lateral displacement. *Science*, 304, 987–990.

Kamholz, A. E., Weigl, B. H., Finlayson, B. A., and Yager, P. (1999). Quantitative analysis of molecular interaction in a microfluidic channel: The T-sensor. *Analytical Chemistry*, 71, 5340–5347.

Marre, S. and Jensen, K. F. (2010). Synthesis of micro- and nanostructures in microfluidic systems. *Chemical Society Reviews*, 39, 1183–1202.

Menasveta, M. J. and Hoagland, D. A. (1991). Light-scattering from dilute poly(styrene) solutions in uniaxial elongational flow. *Macromolecules*, 24, 3427–3433.

Perkins, T. T., Smith, D. E., and Chu, S. (1997). Single polymer dynamics in an elongational flow. *Science*, 276, 2016–2021.

Sasso, L., Norgaard, T. K., Snakenborg, D., and Kutter, J. P. (2009). A deterministic lateral displacement device for continuous-flow separation of nanometer-sized particles. *American Biotechnology Laboratory*, 27, 13–15.

Schilling, E. A., Kamholz, A. E., and Yager, P. (2002). Cell lysis and protein extraction in a microfluidic device with detection by a fluorogenic enzyme assay. *Analytical Chemistry*, 74, 1798–1804.

Song, H. and Ismagilov, R. F. (2003). Millisecond kinetics on a microfluidic chip using nanoliters of reagents. *Journal of the American Chemical Society*, 125, 14613–14619.

Squires, T. M. and Quake, S. R. (2005). Microfluidics: Fluid physics at the nanoliter scale. *Reviews of Modern Physics*, 77, 977–1026.

Stroock, A. D., Dertinger, S. K. W., Ajdari, A., Mezic, I., Stone, H. A., and Whitesides, G. M. (2002). Chaotic mixer for microchannels. *Science*, 295, 647–651.

Stroock, A. D. and McGraw, G. J. (2004). Investigation of the staggered herringbone mixer with a simple analytical model. *Philosophical Transactions of the Royal Society of London Series A—Mathematical Physical and Engineering Sciences*, 362, 971–986.

Takagi, J., Yamada, M., Yasuda, M., and Seki, M. (2005). Continuous particle separation in a microchannel having asymmetrically arranged multiple branches. *Lab on a Chip*, 5, 778–784.

Takayama, S., Ostuni, E., Leduc, P., Naruse, K., Ingber, D. E., and Whitesides, G. M. (2001). Laminar flows—Subcellular positioning of small molecules. *Nature*, 411, 1016–1016.

Takayama, S., Ostuni, E., Leduc, P., Naruse, K., Ingber, D. E., and Whitesides, G. M. (2003). Selective chemical treatment of cellular microdomains using multiple laminar streams. *Chemistry and Biology*, 10, 123–130.

Tegenfeldt, J. O., Prinz, C., Cao, H., Chou, S., Reisner, W. W., Riehn, R., Wang, Y. M., Cox, E. C., Sturm, J. C., Silberzan, P., and Austin, R. H. (2004). The dynamics of genomic-length DNA molecules in 100-nm channels. *Proceedings of the National Academy of Sciences of the United States of America*, 101, 10979–10983.

Vig, A. L. and Kristensen, A. (2008). Separation enhancement in pinched flow fractionation. *Applied Physics Letters*, 93, 3.

Wautelet, M. (2001). Scaling laws in the macro-, micro- and nanoworlds. *European Journal of Physics*, 22, 601–611.

Weigl, B. H. and Yager, P. (1999). Microfluidics—Microfluidic diffusion-based separation and detection. *Science*, 283, 346–347.

Yamada, M., Nakashima, M., and Seki, M. (2004). Pinched flow fractionation: Continuous size separation of particles utilizing a laminar flow profile in a pinched microchannel. *Analytical Chemistry*, 76, 5465–5471.

Zheng, B., Gerdts, C. J., and Ismagilov, R. F. (2005). Using nanoliter plugs in microfluidics to facilitate and understand protein crystallization. *Current Opinion in Structural Biology,* 15, 548–555.

Zheng, B., Roach, L. S., and Ismagilov, R. F. (2003). Screening of protein crystallization conditions on a microfluidic chip using nanoliter-size droplets. *Journal of the American Chemical Society,* 125, 11170–11171.

ENDNOTES

1. The transition from laminar fluid motion to turbulent flow is characterized by the Reynolds number, which is a dimensionless number that gives a measure of the ratio of inertial forces to viscous forces.
2. For time varying flows, these curves split up in three types with different properties: streamlines, streaklines, and pathlines.
3. The term *advection* is sometimes confused with the term *convection*, but the following definition from Wikipedia stresses the right use of the word: "*Advection*, in chemistry and engineering, is a transport mechanism of a substance, or a conserved property, by a fluid, due to the fluid's bulk motion in a particular direction."
4. Contour length equals the length of the polymer chain backbone in the fully extended state.
5. This is generally a good approximation when the advecting flow only varies on length scales much larger than the particle diameter.
6. For example, field flow fractionation (FFF) and split-flow thin (SPLITT) (see Yamada et al., 2004).
7. For example, hydrodynamic chromatography (HDC) and capillary hydrodynamic fractionation (CHDF) (see Yamada et al., 2004).
8. Winnie Svendsen, DTU Nanotech, Denmark. Personal communication.
9. Many call it the Einstein–Smoluchowski relation, because of Marian Smoluchowski, who independently of Einstein brought the solution of the problem in 1906.
10. As argued earlier in the section titled "Streamlines and Particle Traces," even 5 µm suspended particles follow the corresponding streamline, which is definitely also true for smaller particles/molecules.
11. The diffusive flux can even work against a flow, as shown in the lower-left picture of Figure 8.9 where the concentration field extends upstream into the clear buffer inlet.
12. The Péclet number is named after the French physicist Jean Claude Eugène Péclet (1793–1857).
13. Readers might question if mixing is an actual manipulation method, but as will be apparent in this section, the mixing of miscible fluids on the micron scale is a nontrivial matter, which furthermore is an important task in many lab-on-a-chip systems.
14. Example taken from the book by Henrik Bruus, *Theoretical Microfluidics*, Oxford master series in condensed matter physics, Oxford University Press (2008).
15. This assumption is valid for shallow channels, where the channel height is considerably smaller than the width of the interdiffusion zone (the diffusive broadening of the interface). See the article by Kamholz et al. (1999).
16. A good theoretical introduction to mixing is given by J. M. Ottino, The *Kinematics of Mixing: Stretching, Chaos, and Transport* (Cambridge University Press, Cambridge, 1989).

Index

T - #0721 - 101024 - C264 - 234/156/12 - PB - 9781138381995 - Gloss Lamination